SECONDARY CERTIFICATE SERIES

Biology

Secondary Certificate Series

General Editor   K. W. Watkins, B.SC.(ECON.), PH.D.

FACING LIFE'S CHALLENGE   A STUDY OF MARK'S GOSPEL FOR TODAY
J. Hills Cotterill, B.A., DIP.TH.

SECONDARY CERTIFICATE MATHEMATICS   D. T. Daniel, B.A.

SECONDARY CERTIFICATE ENGLISH   P. S. Morrell, B.A.

COMMERCE AND LIFE   K. Lambert, B.SC.(ECON.)

PHYSICS   I. M. L. Jenkins, B.SC., and E. J. Hanmore, B.SC.

CHEMISTRY   E. J. Hanmore, B.SC.

RURAL SCIENCE COURSE   J. A. Shorney, DIP.R.ED.

HOUSECRAFT TODAY   G. M. Sutton

LIFE IN OUR SOCIETY   K. Lambert, B.SC(ECON.)

Revised and Metricated Edition
Biology

H. T. PASCOE, B.SC.
Head of Science Department,
Bishop Lonsdale College of Education, Derby

NELSON

THOMAS NELSON AND SONS LTD

36 Park Street London W1Y 4DE
P.O. Box 18123 Nairobi Kenya

THOMAS NELSON (AUSTRALIA) LTD
597 Little Collins Street Melbourne 3000

THOMAS NELSON AND SONS (CANADA) LTD
81 Curlew Drive Don Mills Ontario

THOMAS NELSON (NIGERIA) LTD
P.O. Box 336 Apapa Lagos

THOMAS NELSON AND SONS (SOUTH AFRICA) (PTY) LTD
51 Commissioner Street Johannesburg

© *H. T. Pascoe* 1966
First published 1966
Second edition 1971

ISBN 0 17 438005 4

Cover photograph is an electron-micrograph of an ultra-thin (0·00005 mm.) section of a bean leaf bud, showing parts of three adjoined cells magnified 20,000 ×.
Part of a nucleus bounded by its membranes fills the lower left corner. Mitochondria and immature plastids (organelles of respiration and synthesis), each enclosed by its own double membrane, lie in the cytoplasm, which also contains elaborate membrane-bounded cavities, small vesicles and a host of minute protein-forming particles called ribosomes.

Printed & bound in Great Britain by
Hazell Watson & Viney Ltd., Aylesbury, Bucks

Contents

CHAPTER 1 Plant and animal studies from an evolutionary point of view — 9

Experiment 1 What are the similarities and differences between a human cheek cell and an onion epidermal cell?

Experiment 2 Observation of a sea anemone

Experiment 3 Observation of a snail

Experiment 4 Does a frog use its lungs faster after exercise than before exercise?

Experiment 5 What is the purpose of pollination?

Experiment 6 To discover what happens to pollen grains when they are placed in sugar solutions of different strengths

CHAPTER 2 Food manufacture in green plants — 13

Experiment 1 To find out if the foods we eat contain starch

Experiment 2 To find out if the foods we eat contain the sugar glucose

Experiment 3 What happens to starch and glucose when they are heated?

Experiment 4 To discover if leaves contain starch

Experiment 5 To discover if starch is found in leaves kept in an atmosphere devoid of carbon dioxide

Experiment 6 Is light necessary for the formation of starch in green leaves?

Experiment 7 To discover if a green water plant produces a gas when exposed to light

Experiment 8 The inter-relations of plants and animals

CHAPTER 3 Release of energy from food — 19

Experiment 1 To discover how fast we breathe

Experiment 2 The Model Lung

Experiment 3 To discover the volume of air that we breathe out

Experiment 4 To discover how much air we breathe in

Experiment 5 What happens to a lighted candle when it is placed in breathed out air?

Experiment 6 Is air changed during breathing?

Experiment 7a To discover if animals breathe out carbon dioxide

Experiment 7b To discover if germinating peas produce carbon dioxide

Experiment 8 How much of the air is used up in breathing?

Experiment 9 To discover if temperature changes occur when seeds germinate

Experiment 10 To discover if a mouse is gaining or losing weight while it is breathing

Experiment 11 To discover under what conditions hæmoglobin changes colour (see Chap. 7, Exp. 2)

CHAPTER 4 Food values — 28

Experiment 1 To discover if foods contain starch

Experiment 2 To discover if foods contain glucose

Experiment 3 To discover if foods contain protein

Experiment 4 To discover if foods contain fat

Experiment 5 What happens when we heat a sugar cube or a peanut?

Experiment 6 How much energy can we obtain by burning a peanut or cube of sugar?

Experiment 7 To discover if cereals contain iron

Experiment 8 To find out the percentage of water in foods

CHAPTER 5 Digestion of food in mammals 31

Experiment 1 To discover how our teeth are arranged

Experiment 2 To discover what happens to starch when it is mixed with saliva

Experiment 3 To discover how fats are emulsified

CHAPTER 6 Water and its importance to animals and plants 33

Experiment 1 Will a plant grow in unwatered soil or if only the leaves are watered?

Experiment 2 Is water taken up by the roots of a plant?

Experiment 3 To measure the resistance to bending of plant stems

Experiment 4 What are the main regions of the root of a seedling?

Experiment 5 To discover what happens when a concentrated solution is separated from water by a membrane

Experiment 6 Does any change in weight take place when potato chips are placed in water or in a strong salt solution?

Experiment 7 To discover if a liquid is given out by a plant

Experiment 8 To discover if water is given out by the leaves of a plant

Experiment 9 To discover if a leaf has pores

Experiment 10 Do leaves give out water vapour when they are covered with vaseline?

Experiment 11 Does the water escaping from the leaves of a plant help to draw water up the stem?

Experiment 12 What is the nature of the surface of the skin and what is its reaction to a volatile liquid?

Experiment 13 Are all parts of the hand equally sensitive to heat and cold?

Experiment 14 Are our hands a reliable guide to the temperature of water?

Experiment 15 To find out if all parts of the hand are equally sensitive to touch

CHAPTER 7 The structure, function and circulation of blood in vertebrates 41

Experiment 1 What is the blood composed of?

Experiment 2 To discover under what conditions haemoglobin changes colour.

Experiment 3 To discover what happens to a drop of blood when it is exposed to the air

Experiment 4 In which direction is the blood flowing through the veins?

Experiment 5 To discover how blood circulates through the tail or external gills of a tadpole

Experiment 6 To find out the internal structure of a sheep's heart

Experiment 7 Does the pulse beat vary after exercise?

Experiment 8 How quickly will snails come out of hibernation?

Experiment 9 To discover what external conditions cause snails to hibernate

CHAPTER 8 Animal movement and its control 45

Experiment 1 How does an earthworm move?

Experiment 2 To discover how a snail moves

Experiment 3 How is the lower arm moved?

Experiment 4 To discover what happens to a muscle if it is given a small electric shock

Experiment 5 Testing our reflexes

CHAPTER 9 The senses. 47

Experiment 1 To find out the action of the iris under different intensities of light

Experiment 2 How far from the eye must an object be held for its image to fall on the blind spot?

Experiment 3 Why do we need two eyes?

Experiment 4 To find out if the sensitivity to hearing of members of the class varies

Experiment 5 To find out if sea anemones have a sense of taste

Experiment 6 To find out if earthworms have a sense of taste

Experiment 7 Do our tongues have areas which are more sensitive to some tastes than others?

CHAPTER 10 Chemical co-ordination in plants and animals   49

Experiment 1 To find out the direction of growth of broad bean roots placed vertically and horizontally

Experiment 2 In what region of a broad bean radicle does bending occur?

Experiment 3 To find out if oat seedlings respond to light coming from one direction

Experiment 4 To find out how roots respond to the presence of water

CHAPTER 11 Bacteria and moulds — agents of decay and disease   51

Experiment 1 To discover what happens when food is left uncovered in the air for a short length of time

Experiment 2 To obtain some spores and observe their growth under laboratory conditions

Experiment 3 What is the effect of adding yeast to a sugar solution?

Experiment 4 To discover where bacteria are to be found

Experiment 5 To discover what happens to bacteria in milk when it is heated at different temperatures

Experiment 6 To repeat Pasteur's experiment which proved that bacteria in the air cause decay

Experiment 7 To demonstrate how bacteria can be transferred from place to place

Experiment 8 To discover the effects of antiseptics on the growth of bacteria

Experiment 9 To discover various ways in which food can be preserved

Experiment 10 Do root nodules contain bacteria?

CHAPTER 12 Field studies   56

Experiment 1 To compare the rate of fall of sycamore fruits with those of ash

Experiment 2 To discover how far from the parent plant poppy and delphinium seeds are scattered

Experiment 3 To discover how fast fluffy fruits and seeds fall

Experiment 4 To discover how seeds are dispersed from pods

Experiment 5 To find out what soil is composed of

Experiment 6 How much water does a sample of soil contain?

Experiment 7 Which type of soil holds most water?

Experiment 8 How much air does a sample of soil contain?

Experiment 9 To discover some of the physical properties of soils

Experiment 10 To find out how different coloured soils are affected by heat

Experiment 11 What is the effect of the addition of lime to clay?

Experiment 12 How much humus does a soil contain?

Experiment 13 To find out if a sample of soil contains chalk

Experiment 14 To determine the degree of acidity of a soil

Experiment 15 To investigate the growth of plants in different solutions

Experiment 16 How many worms are to be found in a certain area of soil?

Experiment 17 What is the effect on the soil of harrowing or hoeing?

Experiment 18 How far upwards do the spore cases of Pilobolus travel?

Experiment 19 How do woodlice behave when given the choice of humid or dry conditions?

Experiment 20 Observation of woodlice in a dry and a damp atmosphere

Experiment 21 How do woodlice behave when given the choice of humid or dry conditions at different temperatures?

Experiment 22 How do woodlice behave in relation to light?

When you carry out the experiments suggested in this book, first read through the instructions given under the heading Method, *Make sure that you have all the materials you need and that you know what to do. Often there is a diagram to help you. Ask your teacher about any points that do not seem clear to you.*

Watch carefully whilst you are doing the experiment and then write a few short sentences describing what you saw. The questions under the heading Observations *will show you what to look for particularly and the statements under* Conclusions *will guide your thinking.*

Sometimes you will be able to explain how it was that things happened as they did. This is the way that scientists go to work, understanding more and more about the universe by thinking and talking about results of experiments.

Nowadays science covers a vast range of topics, so it is convenient to divide it into sections. The experiments in this book deal with subjects that are grouped under the heading "Biology".

CHAPTER 1

Plant and animal studies from an evolutionary point of view

EXPERIMENT 1
What are the similarities and differences between a human cheek cell and an onion epidermal cell?

METHOD
1. Take the blunt end of a sterilised scalpel and lightly scrape the inside of the cheek.
2. Place the resulting drop of material on a clean glass slide, add a drop of water, and gently lower on a cover slip.
3. Carefully strip off from an onion scale a piece of the thin covering tissue or epidermis.
4. Place a small piece of this in a drop of water on a glass slide and cover with a cover slip.

Look at both slides, in turn, under the low power of the microscope.

OBSERVATIONS

Can you see any small 'bodies' floating, either singly or in masses, in the water? If so, what is their shape? Does this shape differ in the two slides? If so, in what way? What are these small bodies called? Have they any contents? If so, what are they?

Put a drop of iodine at one side of the cover slip and a piece of filter paper at the opposite side. What happens to the iodine? What changes, if any, take place in the contents of the small bodies?

Draw some of the small bodies (cells), label the parts that you can see, and write down any noticeable differences between the two slides.

EXPERIMENT 2
Observation of a sea anemone

(*Suitable anemones for these experiments include the Opelet Anemone* (Anemonia sulcata) *and the Plumose Anemone* (Metridium senile).)

METHOD

(a) *General Structure*

What colour is the anemone? How would you describe its shape? How many tentacles has it? Does it appear to move about a great deal? If not, in what way is its shape an advantage for a fixed mode of life? Does it appear to have a mouth? If so, where is it? Do you think the body is hollow or solid? How could you find out which it is?

(b) *Feeding responses* (see Chapter 9, Experiment 5)

(c) *Response to touch*

What happens if you touch one of the tentacles with a glass rod? Does all of the animal seem to be affected or only the one tentacle? Does the same thing happen if you touch the area inside the tentacles? What happens if you keep on touching the same area? Does the effect seem to spread? Now touch the body near the foot region. What happens? Is the response the same as that seen in touching the tentacle? If not, how do you account for the difference? How does the behaviour of the animal relate to its way of life? If possible, make sketches to illustrate your observations.

EXPERIMENT 3
Observation of a snail

METHOD

1. Look for a number of long rod-like structures on the head of the snail. How many are there? Exactly how are they placed? Observe how they are used as the snail moves. For instance, does it use all of them in the same way? Touch each of these tentacles gently with the tip of a pencil or a pin. What does the snail do? Can you see a black dot at the very tip of two of the tentacles? What happens to this black dot when you try to touch it with a pin? Place the snail in the dark and shine a light on it. What is the response? What do you think the black dots are? Test all areas of the snail's head, body, and foot for sensitivity to touch.
2. How would you describe the texture of the body? Does it appear to be divided into segments? What do you think is the function of the shell? From a comparison of large and small snails, what do you think the rings on the shell indicate?
3. Place the snail on a glass plate over a piece of paper marked with rings which are drawn at intervals of 5 mm. Arrange the glass so that the snail is at the centre of the scale and measure how far it travels in one minute. Repeat this if possible several times. Is there any variation in the distance travelled? What is the snail's average speed/minute?
4. Place a plastic ruler on a slope and encourage a snail to travel up it. Measure its rate of movement as before. Try to get it to travel downhill so that you can determine whether this pace is different from the uphill rate. Record the results in the form of a table.
5. Use your results to calculate how far the snail would move in an hour. How long would it take to move a kilometre? Do you think a snail could move a kilometre without stopping? If you were walking at a snail's pace, how long would it take you to walk to school from your house?
6. While you tested the snail's pace, could you see *how* it moved? Set up a razor blade 'cage'. Use four single-edge razor blades and a block of wood of approximately the same size so that the sharp edges protrude about 2 mm above the wood. Set the snail in the centre of the cage so that if it moves off the wood it will have to move across the edges of a razor blade. If it does this, what happens to the foot, and why?
7. Place the snail on a glass plate, support a mirror underneath and look for changes in the sole of the snail's foot as it moves. Describe the muscular action. How does it help the snail to move? What is left behind on the surface as the snail moves forward? What is its significance?
8. Place a vertical glass plate in front of a snail which is moving. What does it do? If it starts to move upwards, gently turn

the plate 90°. What does the snail do? Turn the plate 90° in the other direction. Does the snail still make the same response? Why do you think it makes this response? Weigh the snail. When the snail is moving vertically hang weights from the shell by looping a thread round the shell and holding it in place with sellotape. What is the maximum weight the snail can lift before it starts to slide backwards? How many times its own weight can your snail lift? Compare this result with that of other members of the class. Repeat the procedure, only this time with a sheet of glass paper instead of the glass plate. Can your snail still lift the same weight?

9 As the snail moves on the glass examine with a hand lens the T-shaped opening near the front of the foot. Does it look as if the snail has a mouth on its foot? Place a narrow strip of lettuce (2 mm wide) near the front of the snail. Watch for the head to emerge from the shell. If the snail eats some lettuce, describe the movement of the organs used in eating. With a hand lens watch the action of the radula. (This is a little like a tongue with teeth on it.) How does the snail use this in eating the lettuce? Try small pieces of other foods to see if it will eat them.

CONCLUSIONS

From your study of the snail write a description that would help a person to recognise the animal if he had never seen one. Include reports on the way the snail moves and eats, or carries out other important life functions (if you tested others), and the effects of environmental conditions. What are its principal life functions? Has this study left any important ones out? Sketches to illustrate your description will make your 'word picture' more complete.

EXPERIMENT 4

Does a frog use its lungs faster after exercise than before exercise?

METHOD

1 Place a frog in a large beaker containing a little water.
2 Allow the animal to settle down and then count, for exactly half a minute, the number of times its throat moves up and down. Do this three times. Count in the same way the number of times its nostrils twitch.
3 Now exercise the frog in a sink of water for 20 seconds. Return it to the beaker and count its throat and nostril movements as before until they have settled down to normal.
4 Plot your results on a graph showing the number of throat and nostril movements per half minute against time in minutes.

Does the rate of 'throat breathing' increase or decrease after exercise? Does the rate of 'lung breathing' increase or decrease? How do you account for your results?

EXPERIMENT 5

What is the purpose of pollination?

(*Plants suitable for this experiment include daffodil, tulip, bluebell, cherry, lupin, buttercup, snapdragon, foxglove and broom.*)

METHOD

1 Remove all the stamens (before the pollen sacs have burst) from a large number of flowers of one of the above plants. Do this just before the flowers open properly.

2 Divide this group of flowers into two, label one group A and the other B, and cover all the flowers in polythene bags. What animals do you think are now prevented from reaching the flowers?
3 Select a further group of flowers of the same plants, label them C, and leave them untreated.
4 When the stamens of Group C are shedding pollen, remove a few stamens and, having removed the bags from Group B, place some Group C pollen on the central portion of all the Group B flowers. Then replace the bags. Leave for several weeks.

OBSERVATIONS

In which groups have seeds been formed? How many seeds in each group?

What conclusions can you draw from your experiment?

EXPERIMENT 6

To discover what happens to pollen grains when they are placed in sugar solutions of different strengths

METHOD

1 Prepare small quantities of 3%, 5%, 7%, 10% and 15% sucrose solution.
2 Place a drop of each of these solutions on a separate glass cover slip, and to each add a few grains of pollen.
3 Make a number of rings of plasticene on separate glass microscope slides and carefully invert the cover slips over these rings so that the drop of sugar solution hangs down in the cavity so formed.
4 Examine the drops with a lens at intervals. Can you see any tubes growing from the pollen grains? If so, in which solutions? What would be a good name for these tubes?

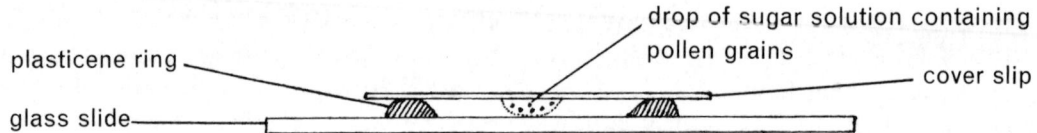

CHAPTER 2

Food manufacture in green plants

EXPERIMENT 1
To find out if the foods we eat contain starch

METHOD

Add a few drops of iodine solution to thin slices of a variety of vegetables and fruits, e.g. potatoes, onions, carrots, cabbage, artichokes, turnips, peas, beans, apples, oranges, bananas, etc.

If the food does contain starch what colour will the starch change to? Does this change occur with any of the foods you are testing?

RESULTS

Indicate your results in the form of a table.

Food	Starch present	Glucose present

Indicate presence with a ✓.

CONCLUSION

.... contain starch.

EXPERIMENT 2
To find out if the foods we eat contain the sugar glucose

METHOD

1 Place a small portion of each of the foods tested in Experiment 1 in a test tube.
2 Add a **few drops** of freshly prepared Fehlings solution and heat the test tube for a few minutes (taking care to hold the mouth of the test tube away from yourself and other pupils).

If the food contains glucose what is the colour change to be seen on heating? Does this change occur with any of the foods you are testing?

RESULTS

Indicate your results in the table that you have already prepared in Experiment 1.

CONCLUSION

.... contain glucose.

EXPERIMENT 3
What happens to starch and glucose when they are heated?

METHOD

1 Place a small quantity of starch or glucose in a dry, hard glass test tube, and fit into it a delivery tube.

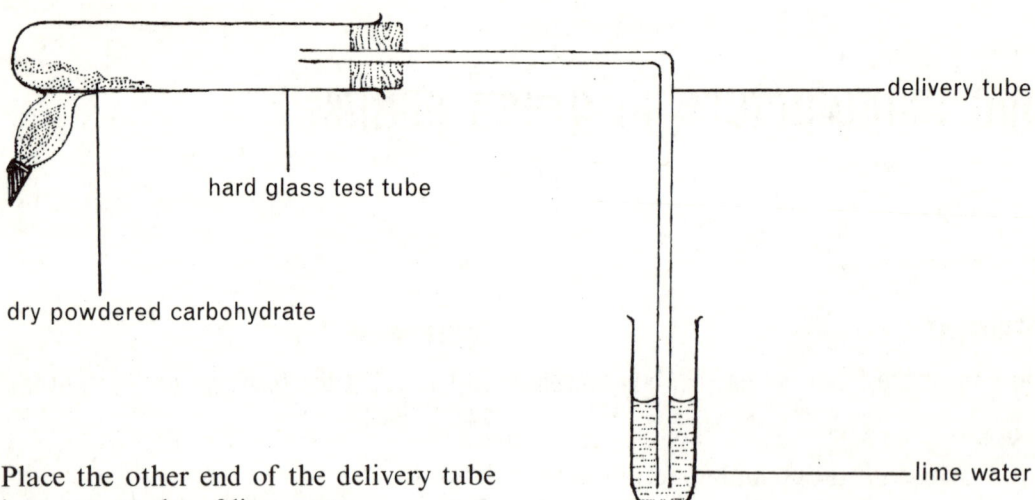

2 Place the other end of the delivery tube into a test tube of lime water.
3 With the hard glass test tube horizontal, heat it with the bunsen burner.
4 After heating for five minutes, remove the tube from the lime water, allow the test tube to cool and remove the cork. Place a small quantity of white anhydrous copper sulphate on the inner surface of the upper part of the tube.

OBSERVATIONS

Do you notice any change taking place in the lime water? Is there any change in the powder after it is placed in contact with the inner surface of the tube? If there is a change, what has caused it? What is the nature of the residue left in the test tube?

CONCLUSIONS

On heating starch or glucose strongly it breaks up into .... and ....

EXPERIMENT 4

To discover if leaves contain starch

METHOD

1 Obtain a number of nasturtium leaves. Divide them into two groups and place the stalks in a beaker of water.
2 Place one group of leaves in a dark cupboard and leave them. Place the other group in the light of a window.
3 After 48 hours, test both groups of leaves for starch as follows:
 (a) Drop the leaves into boiling water for a minute or so. What does this do to the leaves?
 (b) Place the leaves in a beaker half filled with colourless methylated spirits. Float this beaker in a larger one containing nearly boiling water. Make sure the bunsen is turned out. Do you notice any change in the colour of the meths or of the leaves? If the water gets cool quickly, reheat it, but first remove the beaker of meths.
 (c) When the leaves are white remove them from the meths, and drop them into hot water. Leave them for 5 minutes. What is the texture of the leaves on being taken out of the

meths? What is the effect of the hot water on the leaves?

(d) Finally, place the leaves in a dish of iodine solution.

4 Test a variety of leaves, including some variegated ones, as described in 3. Draw and colour the actual shape and appearance of the variegated leaves.

OBSERVATIONS

Do any of the leaves darken or go blue-black in the iodine? If so, which ones? What is the name of the substance which goes blue-black in iodine? Draw and colour the variegated leaves after they have been tested. Which areas correspond to the original white and green areas of the leaf?

CONCLUSIONS

Leaves which have been kept in the light contain ...., whilst leaves kept in the dark .... In the presence of chlorophyll .... is made.

EXPERIMENT 5

To discover if starch is found in leaves kept in an atmosphere devoid of carbon dioxide

METHOD

1 Obtain a number of nasturtium plants which have been kept in the dark.

2 Place 15 mm of soda-lime in the bottom of a large specimen tube, keeping it in place with a plug of cotton wool.

3 Split the cork of the tube into two pieces and make a groove in the two pieces the same size as the stalk of one of the leaves.

4 Holding a leaf stalk firmly between the two pieces of cork, place the leaf into the tube so that the cork fits tightly. Spread vaseline over the outside of the cork.

5 Support the tube with a clamp and stand.

6 Keep the plant well watered and place it in a light place.

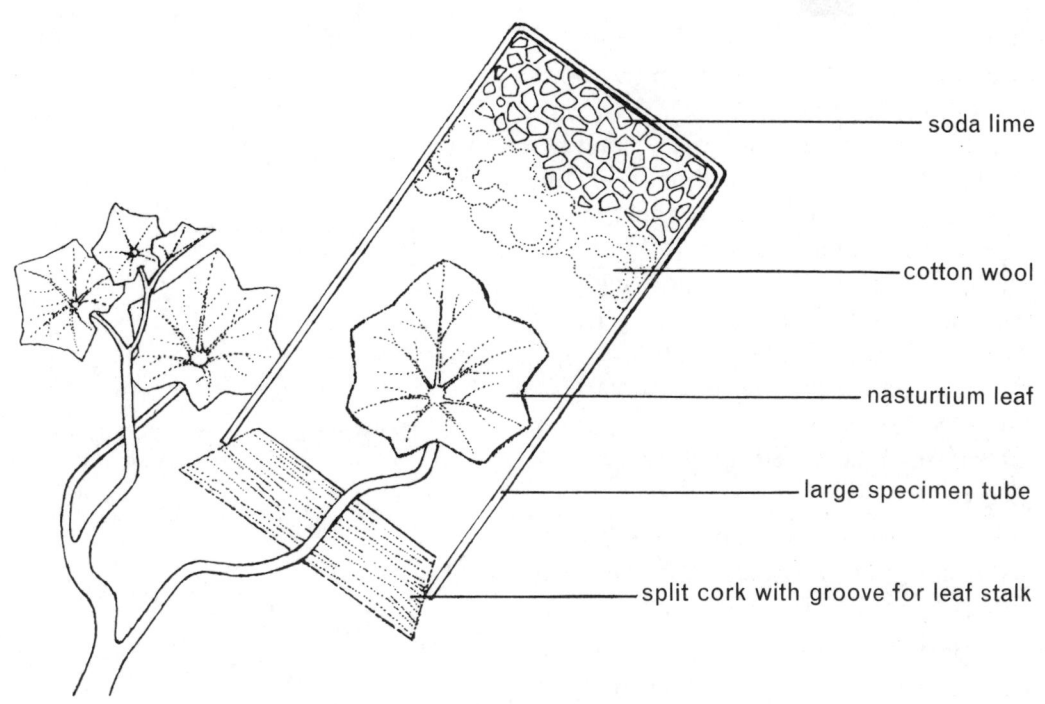

7 After twenty-four hours test both the experimental and a normal leaf for the presence of starch.

OBSERVATIONS

On testing for starch, which leaf contains starch? What is the purpose of the soda lime?

CONCLUSION

Leaves in the absence of carbon dioxide .... starch.

EXPERIMENT 6

Is light necessary for the formation of starch in green leaves?

METHOD

1 Place some nasturtium leaves in the dark for 48 hours.
2 Cut a simple pattern out of the middle of a small square of cardboard or tin foil. Carefully attach this to a leaf using paper clips, with a piece of cardboard or tin foil of the same size covering the underneath of the leaf.
3 Expose to light for a few hours.
4 Test the leaf for the presence of starch.
5 Draw the leaf with its 'stencil' before and after it has been tested for starch.

RESULTS

Where has the starch been formed? Why has the starch been formed there and not in other regions?

CONCLUSION

Leaves need .... in order to manufacture ....

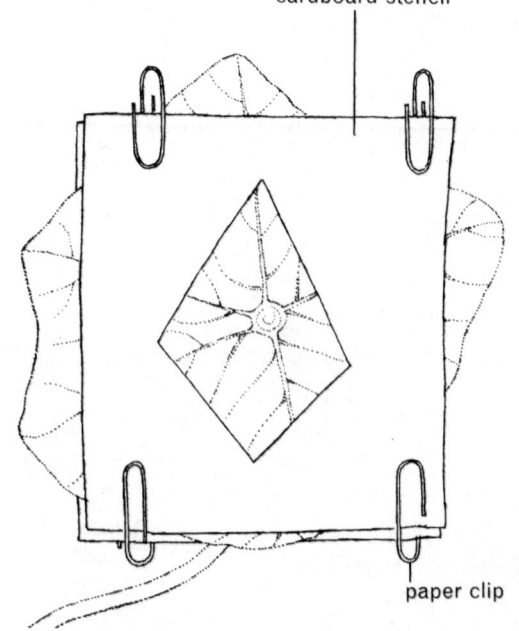

EXPERIMENT 7

To discover if a green water plant produces a gas when exposed to light

METHOD

1. Place a piece of pond weed in the bottom of a beaker.
2. Place the beaker in a sink filled with water.
3. Place a filter funnel in the beaker to cover the pond weed.
4. Submerge a test tube in the water, ensure that it is full of water and without any air bubbles, and place it, inverted, over the 'spout' of the filter funnel.
5. Remove the now completed apparatus from the water and carefully pour away a little of the water from the beaker taking care to ensure that the open end of the test tube stays under the water.
6. Gently breathe into the water for a few minutes through a glass tube.

OBSERVATION

At the end of a few days, what do you notice has happened in the inverted test tube which you had previously filled with water? What has caused this change?

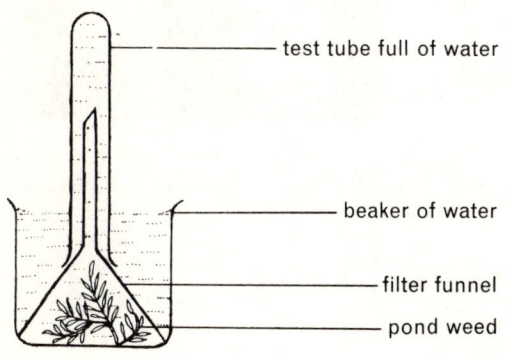

Test any gas which may have collected in the tube as follows. Carefully place the apparatus under sufficient water for the beaker to be just covered. Holding the test tube, remove the rest of the apparatus from underneath it, and put your finger over the open end, trying to trap as little water as possible. Remove the test tube from the water and test the gas with a glowing splint. Record what happens. What is the name of the gas?

CONCLUSION

A green water plant when exposed to the light produces ....

EXPERIMENT 8

The inter-relations of plants and animals

METHOD

1. Each group needs six large specimen tubes. Label these 1 to 6. Set them up as follows:

   In each tube place 15 mm of freshly washed sand and then fill nearly to the top with distilled water.

   Tube 1—leave untouched.
   Tube 2—add one pond snail.
   Tube 3—add two sprigs of pond weed.
   Tube 4—add one pond snail and two sprigs of pond weed.

   Enclose the tops of these four tubes with a piece of polythene held in place by an elastic band. Place them in a well-lit place.

   Tube 5—add one pond snail.
   Tube 6—add two sprigs of pond weed.

   Enclose the tops of these two tubes as above. Place them in the dark.

   Leave all the tubes set up for at least a day.

2 Place a few ml of bromo-thymol blue indicator in a test tube. Make a note of its colour. Now blow gently through a glass tube for a few minutes into the indicator. At the end of this time has the indicator changed colour? If so, what is the change and what do you think might have caused this?

3 Now test the tubes with this indicator as follows: Obtain a glass tube drawn out into a fine point. Suck up into this approximately 2 ml of indicator, and keeping your finger over the end plunge the pointed end of the tube through the polythene cover of the specimen tubes and release the indicator. Place each specimen tube in front of a piece of white paper and make a careful note of the resulting colour of the water.

RESULTS

Tube No.	Colour of water
1	
2	
3	
4	
5	
6	

CONCLUSIONS

Keeping in mind what happens when you blow into the indicator, what possible conclusions can you draw to account for your results?

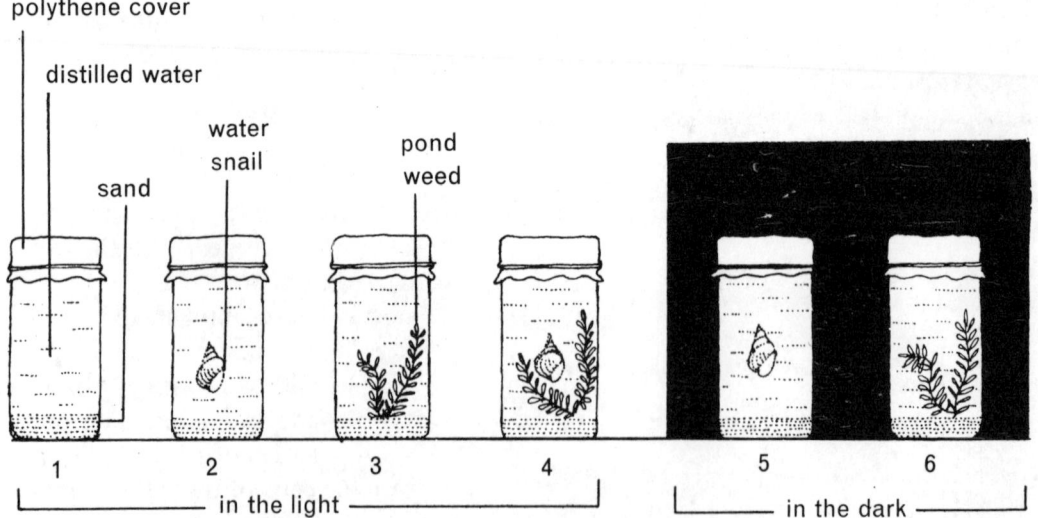

CHAPTER 3

Release of energy from food

EXPERIMENT 1
To discover how fast we breathe
METHOD

1. Work in pairs and watch the breathing movements of your partner's chest.
2. For one minute count the number of times your partner breathes in.
3. Repeat this for another minute and then take the average of your two results. This number will be the breathing rate/min.
4. Repeat the experiment after taking exercise such as running up the stairs or once round the school.

RESULTS

Sitting	Breathing rate/min.
Self . .	
Partner . .	
After Running	**Breathing rate/min.**
Self . .	
Partner . .	

Can you account for any differences in the two sets of results?

EXPERIMENT 2
The Model Lung
METHOD

1. Fix up the apparatus as shown in the diagram, making sure that all the connections are airtight.

OBSERVATIONS

What happens to the balloons when the rubber 'diaphragm' is pulled up and down? Why does this happen? Place a piece of smoking paper over the 'mouth'. What happens to the smoke as the diaphragm is pulled up and down?

CONCLUSION

When the diaphragm moves down air is .... and when it moves up air is ....

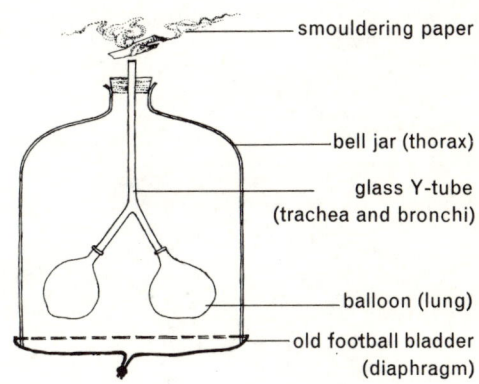

- smouldering paper
- bell jar (thorax)
- glass Y-tube (trachea and bronchi)
- balloon (lung)
- old football bladder (diaphragm)

EXPERIMENT 3a

To discover the volume of air that we breathe out

METHOD

1. Carefully fill a large bottle with water.
2. Fill a trough or the sink nearly to the brim with water.
3. Place the palm of your hand tightly over the open end to prevent any water coming out and with the help of your partner carefully turn the bottle upside down with the open end under the water. What happens to the water in the bottle?
4. Rest the open end on a beehive shelf, one person holding the bottle.
5. Place a length of rubber tubing through the open end into the bottle.
6. Breathe out normally through the rubber tube and into the bottle. What happens?
7. Cork the bottle under water and stand it upright.
8. Using a measuring cylinder find out how much water is necessary to fill the bottle. This volume of water is the volume of air that you breathe out. Why?

RESULT

Volume of water necessary to fill the bottle = ... ml

CONCLUSION

∴ Volume of air breathed out normally = ... ml

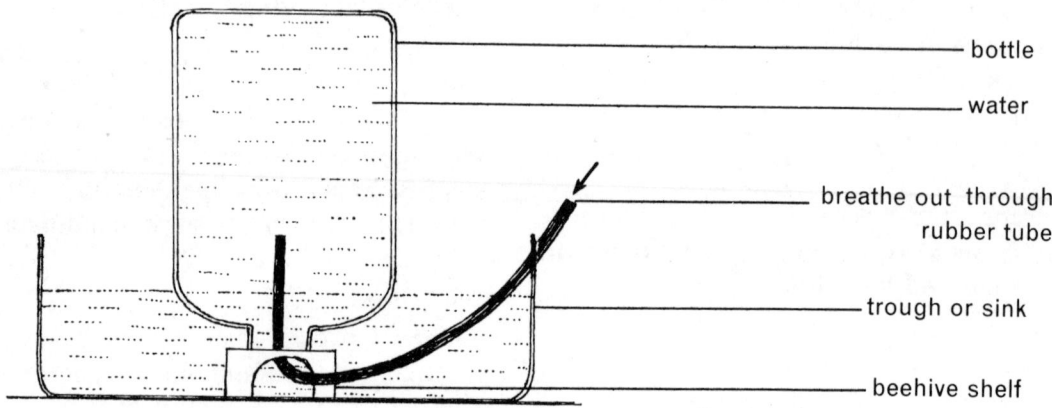

EXPERIMENT 3b

METHOD

1. Repeat the experiment as before but this time take a breath as deep as possible before breathing out into the bottle.
2. Measure how much water must be poured in to fill the bottle, as you did before.

RESULT

Volume of water necessary to fill the bottle = ... ml

CONCLUSION

∴ Volume of air breathed out with a deep breath = ... ml

What is the value of deep breathing?

EXPERIMENT 4

To discover how much air we breathe in

METHOD

1. Use the same apparatus as in Experiment 2 but this time stand the bottle **empty** upside down in the water.
2. Push the rubber tube well up into the bottle and take one normal breath in. What happens as you do this? Why does this happen?
3. Cork the bottle under water, remove it from the water and measure the amount of water required to fill the bottle.

RESULTS

Volume of bottle full of water = ... ml

Volume of water needed to fill the bottle = ... ml

∴ Volume of water that replaces air breathed in = ... ml

CONCLUSION

Volume of air breathed in = ... ml

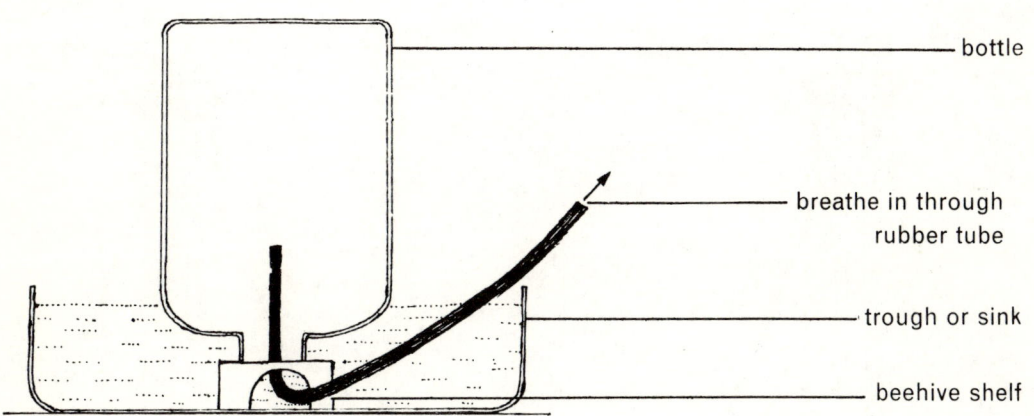

- bottle
- breathe in through rubber tube
- trough or sink
- beehive shelf

EXPERIMENT 5

What happens to a lighted candle when it is placed in breathed-out air?

METHOD

1. Place a gas jar over a small lighted candle. How long does it take to go out? Why does it go out?
2. Collect a gas jar of breathed out air, of the same volume as above, by displacement over water.
3. Quickly remove the gas jar plate and place the jar over the same candle freshly lit.

How long does it take to go out this time? Does it take more or less time than in fresh air? What is the name of the gas that supports combustion? From the results of your experiment would you say that breathed out air contained more or less of this gas than fresh air?

EXPERIMENT 6

Is air changed during breathing?

METHOD

1. Draw air slowly through flask A by sucking at tube M, and time how long it takes for the lime water to turn cloudy.
2. Next breathe out slowly through flask B at tube M, and time how long it takes for the lime water to become as cloudy as the lime water in flask A. Shake flask A gently as you do so.

RESULTS

Time taken for flask A
to become cloudy as
air was breathed in   = ... min ... s

Time taken for flask B
to become cloudy as
air was breathed out   = ... min ... s

CONCLUSION

Air breathed out differs from air breathed in. It contains more of a gas which turns lime water .... This gas is ....

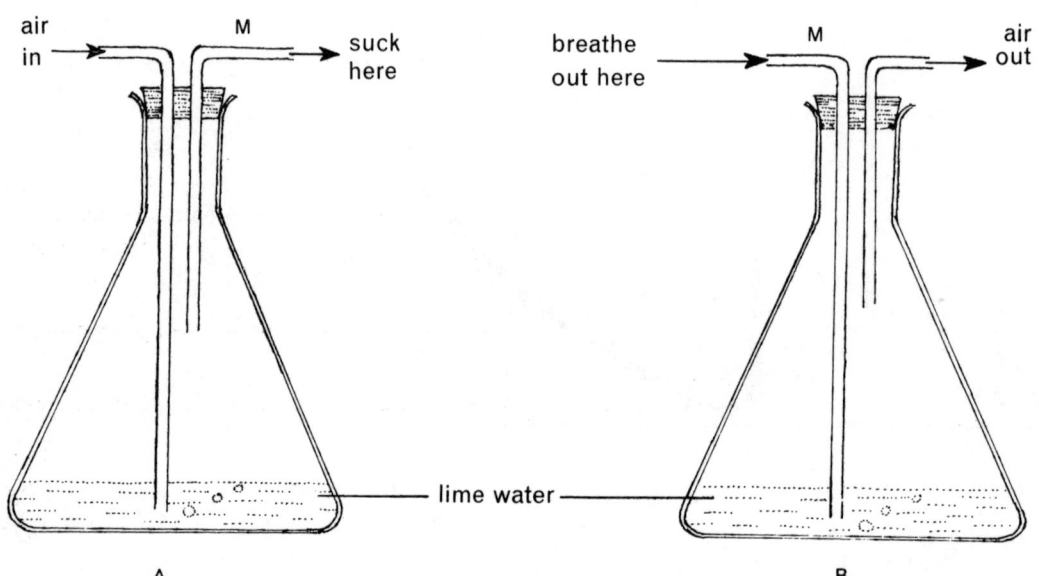

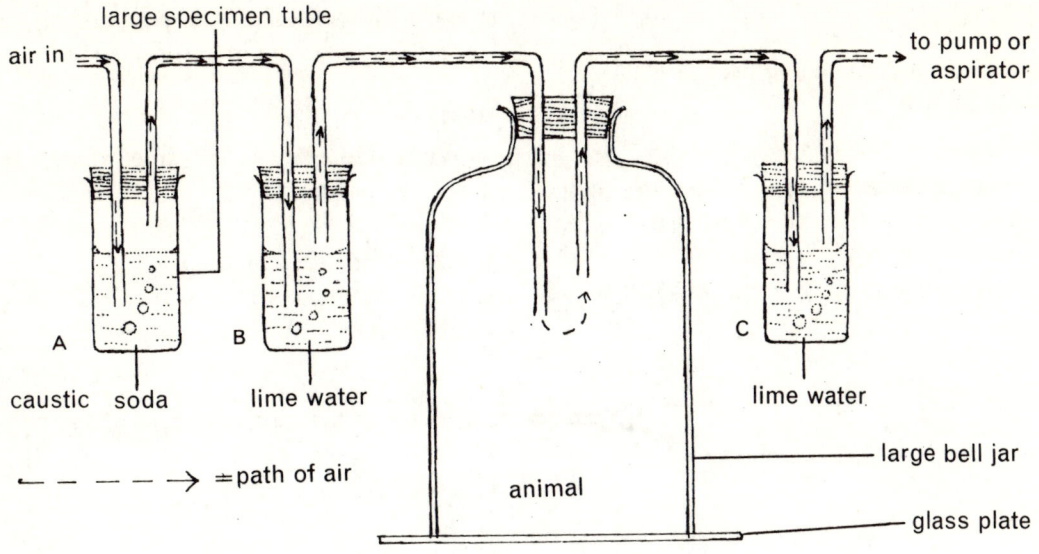

EXPERIMENT 7a

To discover if animals breathe out carbon dioxide

METHOD

1. Place enough caustic soda solution in specimen tube A, and enough lime water in specimen tubes B and C to cover the ends of the longer tubes.
2. Place your animal in the jar (with bedding if necessary).
3. Draw air slowly through the tubes by attaching tube C to a filter pump or aspirator.

What is the purpose of the caustic soda solution?

OBSERVATION

What happens to the lime water in tubes B and C?

CONCLUSION

Account for any changes seen.

EXPERIMENT 7b

To discover if germinating peas produce carbon dioxide

METHOD

1. Place germinating peas in a jar connected by a delivery tube to a test tube of lime water.
2. After a few days attach a football pump to a rubber tube A, open the clip and slowly pump air into the jar.

OBSERVATION

What happens to the lime water? Why does this happen?

CONCLUSION

Germinating peas ....

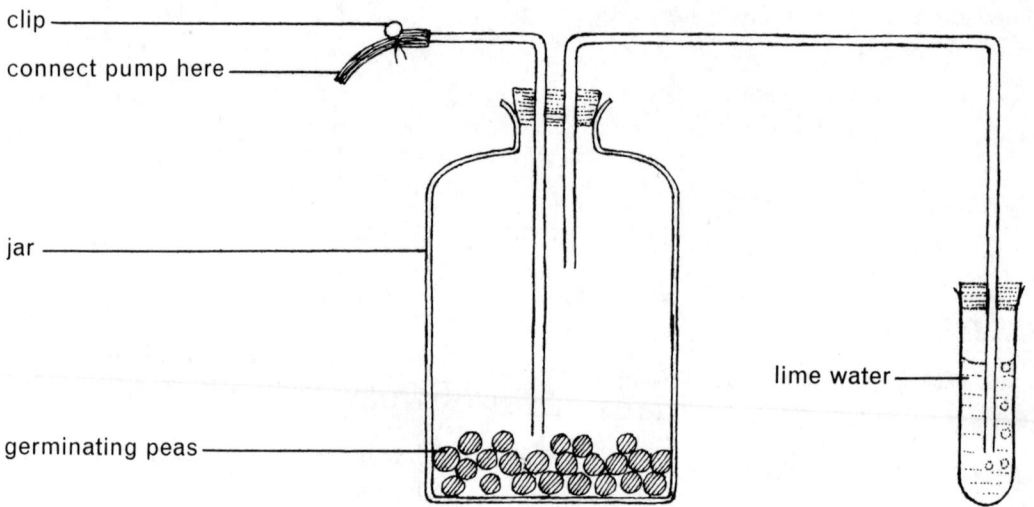

EXPERIMENT 8

How much of the air is used up in breathing?

METHOD
1. Measure the volume of U-tube A.
2. Set up two U-tubes as shown in the diagram.
3. Place the organisms being tested in the corked limb of tube A, and place the open end in a beaker of dilute caustic soda solution.
4. Set up tube B as in 2, but place the open end of the tube in a beaker of water.
5. Measure the levels of the liquids in tubes A and B.

OBSERVATION
Does any change take place in these levels? If so, in which tube? Measure any changes in volume in tube A. How do you account for these? What is the purpose of tube B?

CONCLUSION
Approximately……………of the air has been used up in breathing by………………
This fraction is the gas …………………

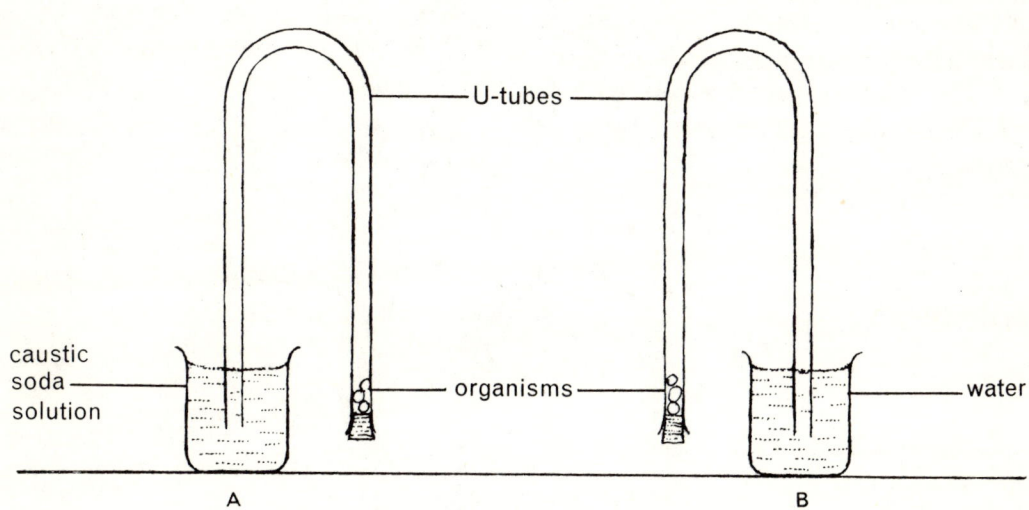

EXPERIMENT 9

To discover if temperature changes occur when seeds germinate

METHOD

1. Place a known weight of soaked peas in thermos flask A, and with a cotton wool plug carefully insert a thermometer. Make sure you have taken enough peas to cover the bulb, and that you can see enough of the thermometer to be able to read the temperature.
2. In flask B place the same weight of peas that have first been soaked in water and then boiled for five minutes. Insert a thermometer as before.
3. Record the temperatures of the two flasks at 9 a.m., 12 noon and 4 p.m. each day for several days. Plot the results as a graph.

RESULTS

Time	9 a.m.	12 noon	4 p.m.	9 a.m.	12 noon	4 p.m.	9 a.m.	12 noon	4 p.m.
Temperature in flask A in °C.									
Temperature in flask B in °C.									

Why have the peas in flask B been boiled for five minutes before being placed in the flask?

Why would it be incorrect to omit flask B from this experiment?

CONCLUSION

The germination of the seeds was accompanied by ....

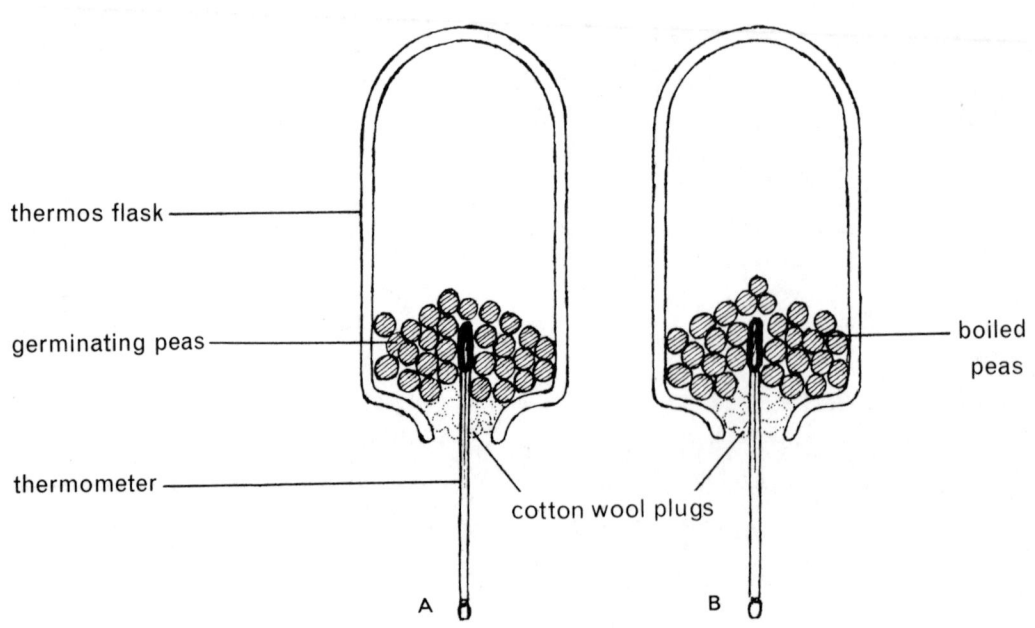

EXPERIMENT 10

To discover if a mouse is gaining or losing weight while it is breathing

METHOD

1 Carefully balance a mouse, in its cage, on one arm of a sensitive balance.

OBSERVATION

Is the mouse gaining or losing weight?

CONCLUSION

How do you account for your observations?

EXPERIMENT 11

To discover when haemoglobin changes colour

METHOD

1 Obtain a few drops of blood as follows: Tie a piece of cotton or string tightly round the last joint of the index finger.
2 Sterilize the skin just below the nail by smearing it with alcohol.
3 Sterilize a needle passing it two or three times through a flame.
4 Bend the end of the finger and prick the skin just below the nail.
What happens? What is the purpose of the tightly tied cotton or string?
5 When you have obtained a few drops of blood, transfer them as quickly as possible to a test tube.
6 Add to the test tube for every drop of blood twenty drops of water.
7 Put your finger over the end of the test tube and shake it vigorously. What is the colour of the liquid?
8 Add a few crystals of sodium hyposulphite. Is there any change in the colour of the liquid? If so, what change?
9 Again, shake the liquid for a few minutes. Is there any change now in the colour of the liquid?

How do you account for your observations?

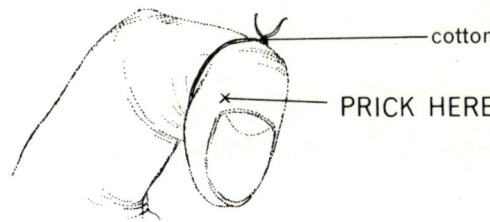

CHAPTER 4

Food values

EXPERIMENT 1
To discover if foods contain starch

METHOD
1 Add a few drops of iodine solution to samples of different foods. If the food does contain starch what colour will the starch change to? Does this change occur with all of the foods you have tested?

RESULTS ....

Indicate your results in a table like the one at the top of the facing page.

EXPERIMENT 2
To discover if foods contain glucose

METHOD
1 Place a small quantity of food in a test tube.
2 Add a **few drops** of Fehlings solution and heat the test tube for a few minutes. (Take care to hold the mouth of the test tube away from yourself and other pupils.)

If the food does contain glucose what colour change is seen on heating? Does this change occur with any of the foods you are testing?

RESULTS

Indicate your results in the table that you have already prepared.

CONCLUSION

....................contain glucose.

EXPERIMENT 3
To discover if foods contain protein

METHOD
1 Add 5 ml of the reagent to the food in a test tube.
2 Warm the mixture.
   The reagent is pale blue. What colour does it change to if protein is present? Do any of the foods being tested show this change?

RESULTS

Indicate your results in the table.

CONCLUSION

....................contain protein.

EXPERIMENT 4
To discover if foods contain fat

METHOD
1 Rub or crush the food on a piece of white paper, and leave for five minutes. Then hold the paper up to the light. Does the area where the food has been allow light to show through? What does this mean?

RESULTS

Indicate your results in the table.

CONCLUSION

....................contain fat.

Food	Starch present	Glucose present	Fat present	Protein present

CONCLUSION

.... contain starch.

EXPERIMENT 5

What happens when we heat a sugar cube or a peanut?

METHOD

1 Fix a sugar cube or a shelled peanut on a needle. Try and light it with a match. What happens?
2 Repeat, only this time dip the sugar cube or peanut in some ashes. Can you light it with a match now? What is the function of the ash?

OBSERVATION

What is produced when these objects burn?

CONCLUSION

When a sugar cube or peanut is burnt in air,...................and.................... are produced.

EXPERIMENT 6

How much energy can we obtain by burning a peanut or cube of sugar?

METHOD

1 Obtain a tin can which is open at one end. Punch holes in the sides and the end. Line the can with asbestos wool.
2 Mount the peanut or sugar cube on a needle set in a cork.
3 Suspend a small test tube containing

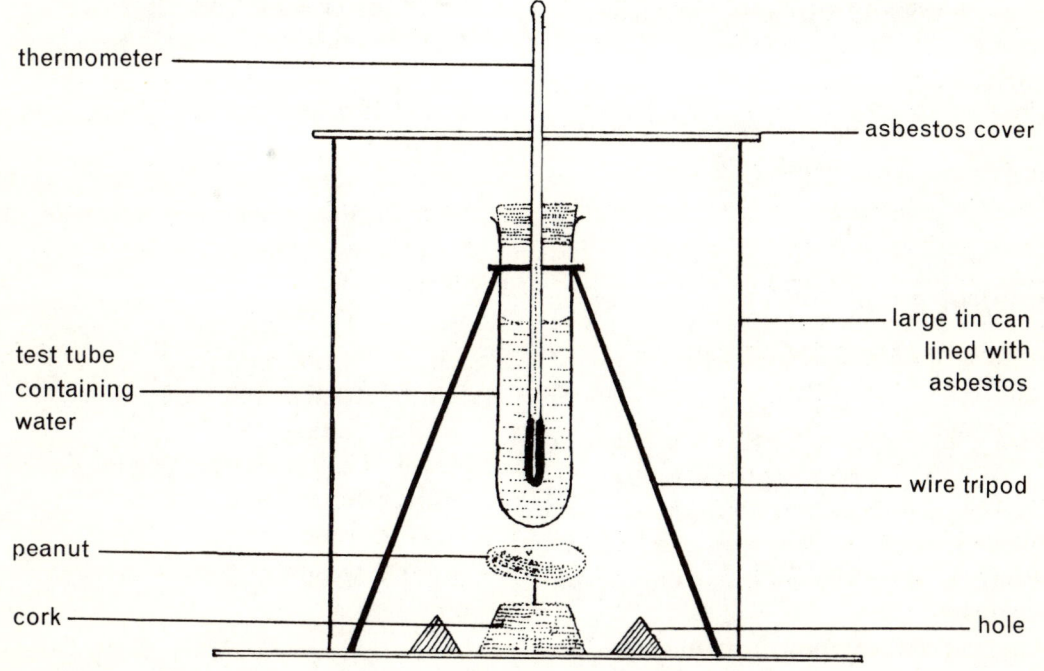

1 ml of water above the peanut or sugar cube by means of a wire tripod.
4 Insert through the corked end of the test tube a Celsius thermometer, and record the temperature of the water.
5 Place a piece of asbestos over the open end of the can. What is the purpose of this and the asbestos lining?
6 Ignite the peanut or sugar cube and when these are completely burnt out, take the temperature of the water again.

RESULTS

Temperature of 1 ml of water
  before heating   = °C
Temperature of 1 ml of water
  after heating   = °C
∴ rise in temperature   = °C

Since 1 ml of water weighs 1 g and the amount of heat required to raise the temperature of 1 g of water 1° C is known as a calorie, the amount of heat energy produced on burning a peanut or sugar cube = (1 × rise in temperature of water) calories. (1 calorie = 4·18 joules.)

Can you think of any ways in which this experiment could be made more efficient?

CONCLUSION

A peanut or sugar cube when burnt produced .... calories.

EXPERIMENT 7

To discover if cereals contain iron

METHOD

1 Burn a few pieces of cereal on a crucible lid until the cereal has turned to ash.
2 Cool the ash and grind it into a powder.
3 Place this powder in a test tube, add a few drops of hydrochloric acid and warm the tube.
4 Filter off the solution from the residue and add to the solution a few drops of $\frac{1}{2}$% potassium ferrocyanide solution.

OBSERVATION

The formation of a blue colour or precipitate indicates the presence of an iron salt. Do any of the cereals you have tested contain iron?

CONCLUSION

.....................contain iron.

EXPERIMENT 8

To find out the percentage of water in foods

METHOD

1 Obtain small pieces of foods such as cabbage, bread, potato and apple, and some peas or beans. Chop them up finely.
2 Take enough of the finely chopped food to fill a crucible.
3 Weigh the crucible and contents.
4 Heat the crucible on a sand bath using a very small flame so as not to char the food. (Or alternatively dry the food in an oven.)
5 After an hour or so allow the crucible to cool, and when cool reweigh the crucible and contents.

RESULTS

Weight of crucible empty   = a g
Weight of crucible and food   = b g
Weight of crucible and food
  after heating   = c g
∴ loss in weight due to heating = $b-c$ g
∴ % of water in food =
$$\frac{\text{loss in weight} \times 100}{\text{original weight of food}} = \frac{b-c}{b-a} \times 100$$

CONCLUSION

.... contains ......% of water.

CHAPTER 5

Digestion of food in mammals

EXPERIMENT 1
To discover how our teeth are arranged

METHOD
1. Draw two semi-circles on your paper. These represent your upper and lower jaws.
2. The following symbols can be used to represent your teeth:
   - Molars — ⊞
   - Premolars — ◊
   - Incisors — ∪
   - Canines — △
3. Open your mouth, and using a small mirror note the position of your teeth. In the semi-circle you have drawn mark their position using the symbols as listed above.
4. When you have completed your drawings, count the number of different teeth and fill in the chart below.

RESULTS

Symbols	U.J.	L.J.
⊞		
◊		
∪		
△		

What is your total number of teeth? How many have fillings? Why do you have differently shaped teeth? How many teeth should you have when you are adult?

EXPERIMENT 2
To discover what happens to starch when it is mixed with saliva

METHOD 1
1. Slowly chew a piece of bread or potato. Do you notice any particular taste developing?

METHOD 2
1. Take 5 ml of starch suspension and pour into four test tubes.
2. Place the test tubes in a beaker of water kept between 35° C and 40° C.
3. Obtain a sample of saliva by swilling a few ml of warm water round your mouth.
4. Divide the saliva into three equal portions in three fresh test tubes.
5. Add one portion of saliva to one test tube of starch and call this test tube A. Place it in the beaker of water.

31

6 Boil another portion of saliva for 5 minutes, add this to a test tube of starch and call this test tube B. Place it in the beaker of water.
7 Add the third portion of saliva to a test tube of starch, add a few ml of vinegar and call this test tube C. Place it in the beaker of water.
8 To the last test tube of starch add an equal volume of water, call this test tube D and place it in the beaker of water.
9 After 15 minutes, take small samples from each test tube in turn and test for the presence of starch with iodine solution and the presence of sugar with Fehlings solution. Record your results in the table below.

RESULTS

Test	A	B	C	D
	starch saliva	starch boiled saliva	starch vinegar saliva	starch water
Iodine				
Fehlings soln.				

What is the purpose of test tube D?

CONCLUSION

When saliva is added to starch, the starch is......................, but if the saliva is heated the starch........................., and when saliva is made acid by vinegar the starch..

EXPERIMENT 3

To discover how fats are emulsified

METHOD

1 Place 5 ml of water in a test tube and add 3 or 4 drops of olive oil.
2 Shake the tube thoroughly. Do the water and oil mix together?
3 Now add **a few drops** of *dilute* caustic soda to the tube.
4 Shake the tube thoroughly. Do the water and oil mix together? How would you describe the appearance of the liquids in the tube?
5 Repeat the experiment as before but this time use liquid paraffin. Do the water and oil mix together?

How would you explain any differences in the results seen with the two oils?

CHAPTER 6

Water and its importance to animals and plants

EXPERIMENT 1
Will a plant grow in unwatered soil or if only the leaves are watered?

METHOD
1. Obtain three weed plants, such as groundsel, from the garden.
2. Put one in a pot with moist garden soil. Call this pot C.
3. Put another plant in a pot with dry soil. Call this pot E.1.
4. Put the other plant in a pot of dry soil and enclose the pot only in a plastic bag. Call this one E.2.
5. Keep pot C well watered (by placing it in a bowl of water). Leave pot E.1 unwatered, and spray water over the leaves of the plant in pot E.2 at regular intervals.
6. Observe the plants day by day.

OBSERVATION

After a few days, do you notice any difference in the state of the plants? What is the difference, if any, after a week? What is the purpose of having the plant in pot C for this experiment?

CONCLUSION

After a week the plant in pot E.1 is.........................., that in E.2......... ...., and that in pot C............ .... The plant in pot C has been used as a.......................................

EXPERIMENT 2
Is water taken up by the roots of a plant?

METHOD A

(*Suitable plants for this experiment include groundsel, dandelion and white deadnettle.*)

1. Carefully dig up a plant and quickly place the roots in a bottle full of water, supporting the stem with a plug of plasticene.
2. Mark the level of water in the bottle with a piece of sticky paper.
3. Set up a similar bottle *without a plant* but with a plug of plasticene and with water up to the same level.

OBSERVATION

After a few days do you notice any change in the levels of water in each bottle? If so, in which bottle has the change taken place? What is the purpose of the second bottle? How do you account for your observations?

CONCLUSION

The roots of a plant........................ ....water.

METHOD B

(*Suitable plants for this experiment include plants with tap roots e.g. dandelion, and plants with white flowers e.g. white deadnettle.*)

1. Carefully dig up two plants making sure

33

that the roots are not damaged. Gently wash the roots in water.

2 Place the roots in a jam jar or bottle of water coloured with eosin (a red dye), and support the stems with a retort stand.

3 After an hour or so carefully make a cut lengthwise through the main root and the stem of one, and cuts across the stem and root of the other. Can you see any trace of the red colour in your sections? If so, make a drawing to show the extent of this. Is any colour to be seen in the leaves (or flowers if they are present)?

How do you account for your observations?

EXPERIMENT 3

To measure the resistance to bending of plant stems

METHOD

1 Take a one foot length of the stem of an annual, or a herbaceous or a woody perennial plant.

2 Secure this length in a vice or retort stand according to the nature of the stem.

3 Attach the hook of a spring balance to the top.

4 Pull slowly on the spring balance at right angles to the stem. Note the reading of the balance when the stem snaps or folds up. (This is a measure of the breaking strain or resistance to bending of the plant.)

5 Repeat for a wide variety of plants. It is important to obtain stems of approximately the same diameter. Why is this? Which are the strongest and which the weakest stems?

6 Record also the angle to the vertical (this measures relative resistance to the wind) when the stem breaks or bends, and measure the height of the plant or tree. Plot a graph of these two measurements. Is there any relation between them?

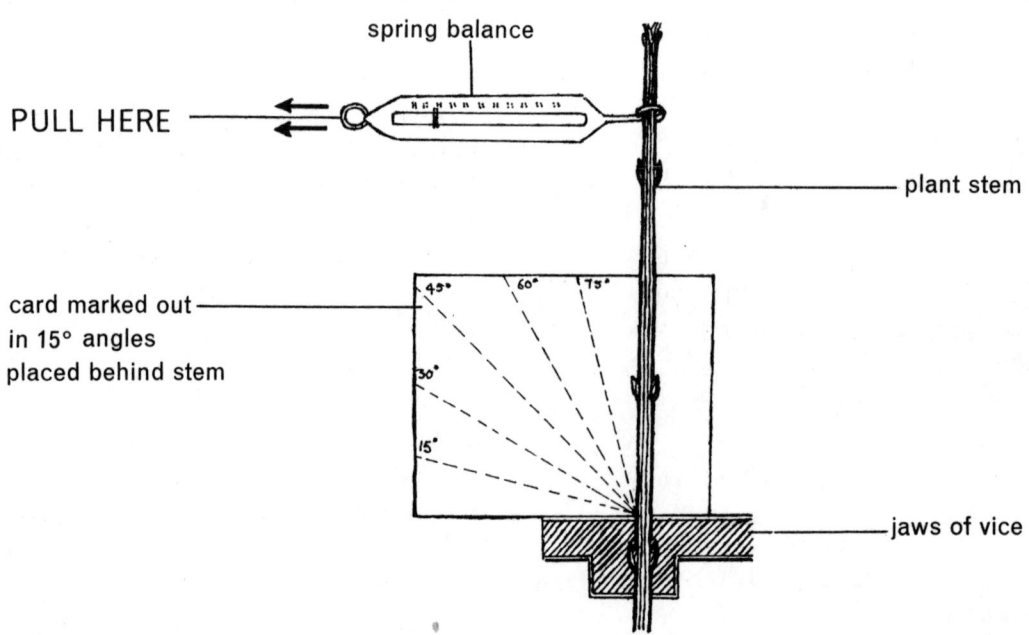

EXPERIMENT 4

What are the main regions of the root of a seedling?

METHOD

1. Germinate cress seedlings by making a 'germination sandwich'. This is made by laying the seeds on the top end of a piece of blotting paper placed on a layer of damp cotton wool. Place one rectangular piece of glass on top of the seeds and one underneath the cotton wool. Hold the glass rectangles together with elastic bands.
2. Place the 'sandwich' with the seeds upwards in a dish containing an inch or two of water. Put the dish in a dark place.
3. After a few days look at the seeds. Have they germinated? If so, look at them carefully. Can you see a mass of white fluffy 'hairs' at one end? If so, which end is this? Are they at the very end? What would be a good name for these hairs? What is their function? Is the tip of the root protected in any way?
4. Make a drawing of a seedling from the 'sandwich', and point out on your drawing those features you have observed above.

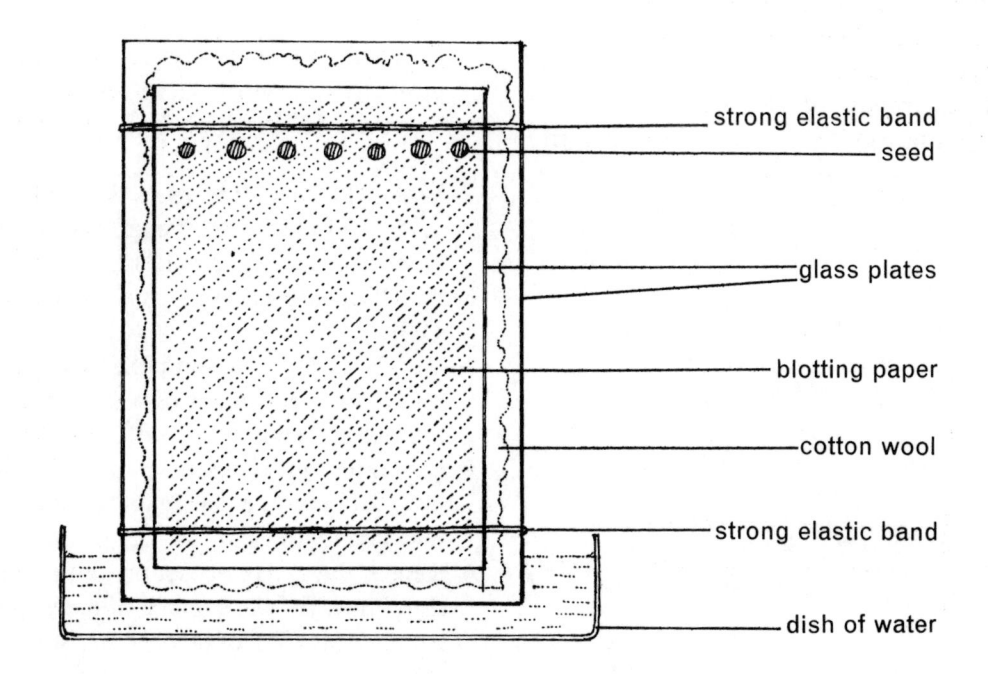

EXPERIMENT 5a

To discover what happens when a concentrated solution is separated from water by a membrane

METHOD

1. Firmly tie on to the broad end of a thistle funnel a piece of pig's bladder or cellophane.
2. Pour into the funnel a strong sugar solution or syrup. This is done either by attaching a filter funnel to the open end and pouring the sugar solution or syrup into this (assisted by a thin piece of wire moved up and down), or by warming the thistle funnel gently with the open end in the liquid. As the funnel cools down the liquid is drawn up into it and it may then be shaken down into the bowl end. This may be repeated a number of times.
3. When enough of the liquid has been poured in to fill the bowl, place this into a beaker of water so that the water covers it and hold the funnel off the bottom of the beaker with a clamp.
4. Mark the level of liquid in the funnel with a piece of paper stuck onto it or with a wax pencil.
5. After a few hours has any change taken place in the level of liquid in the funnel? How do you account for your results?

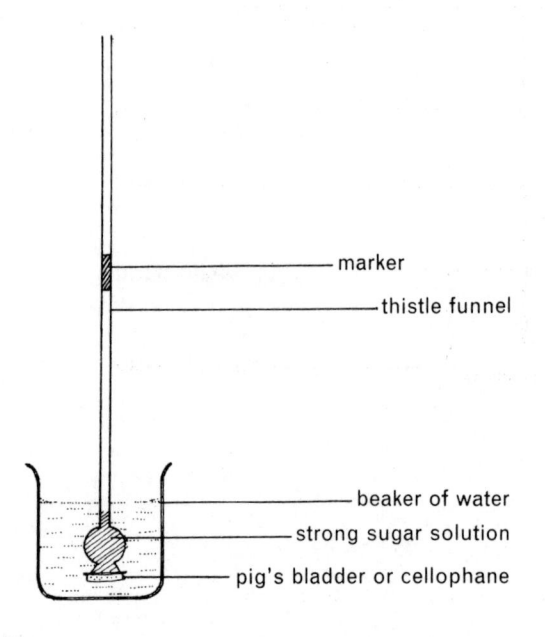

EXPERIMENT 5b

METHOD

1. Carefully place an egg in a beaker of dilute hydrochloric acid. What do you observe? Is anything happening to the shell?
2. When the reaction has ceased, add fresh acid and turn the egg occasionally. Continue to do this until all the shell has disappeared. What is underneath the shell?
3. Gently and slowly pour the egg and acid into a large container of water to remove the acid. Replace the water in the large container with fresh water and with a jam jar remove the egg so that it is floating in the jam jar of water.
4. Repeat the above procedure with another egg but this time place the de-shelled egg in a jam jar of strong salt solution.
5. Make a drawing to show the exact size and shape of the eggs in the two liquids.
6. After a few hours compare the eggs with these drawings. Has any change in size or shape taken place? If so, how do you explain this?

- marker
- thistle funnel
- beaker of water
- strong sugar solution
- pig's bladder or cellophane

EXPERIMENT 6

Does any change in weight take place when potato chips are placed in water or in a strong salt solution?

METHOD

1. Peel a large potato and cut it up into chips about 15 mm wide and 60 mm long. Divide them into two groups.
2. Put each group of chips into a weighed beaker and weigh them.
3. Put one group into a beaker of water and the other into a beaker of saturated salt solution.
4. After a few hours pour off the liquids, quickly and carefully dry the chips, and reweigh them. Do you notice any difference in texture between the two groups of chips?

RESULTS

Group A in Water		Group B in Strong Salt Solution	
Before treatment	g	Before treatment	g
After treatment	g	After treatment	g
Gain or loss in weight		Gain or loss in weight	
Texture		Texture	

CONCLUSION

Potato chips when placed in water .... weight and become ...., while chips placed in strong salt solution .... weight and become .... This is due to ....

EXPERIMENT 7

To discover if a liquid is given out by a plant

METHOD

1. Take a potted plant and completely enclose the pot and the base of the stem in a polythene bag.
2. Place the plant under a dry bell jar.
3. Set up an identical apparatus but without the plant.

OBSERVATION

After a few hours do you notice any change in the inside of the bell jars? If so, what do you think this is due to? Place a little white anhydrous copper sulphate on the inside of the jars. Do you notice any change in colour in the sulphate? If so, how do you account for it? What is the purpose of the apparatus in the second bell jar?

CONCLUSION

What conclusion can you draw from your observations?

EXPERIMENT 8

To discover if water is given out by the leaves of a plant

METHOD

1 Place a leafy shoot in a test tube of water and mark the level of water with a piece of sticky paper or a wax pencil. Place a few drops of olive oil on the surface of the water. Suspend the test tube from the left arm of a balance.
2 Set up an identical apparatus but this time remove all the leaves from the shoot. Suspend this test tube from the right arm of the balance and if necessary add weights to the right pan to obtain a balance between the two tubes.
3 Next day look at the level of water in the two tubes.

RESULTS

Has any change taken place in the level of water in the two tubes? Are the two tubes now balanced? If not, add weights to restore the balance. How much water must be added to achieve the original water levels in any of the tubes?

Weights added to.........-hand
  pan to restore balance   =......g
Volume of water added to
  ....-hand tube to restore
  original water level   =..... ml

CONCLUSION

What conclusions can you draw from your observations and results?

EXPERIMENT 9

To discover if a leaf has pores

METHOD

1 Strip off a thin piece of the outer 'skin' (epidermis) of an iris leaf.
2 Place the piece in a drop of water on a glass slide.
3 Carefully put on a cover slip.
4 Focus the piece under the microscope.

OBSERVATION

Can you see any sausage-shaped cells? If so, how are they arranged and what is between them? What other shapes of cell can you see? Draw the differently shaped cells.

EXPERIMENT 10

Do leaves give out water vapour when they are covered with vaseline?

METHOD

1 Take a number of leaves from a tree and thinly cover with vaseline the upper surface of some and the under surface of others. Place some vaseline on the cut end of the leaf stalk.
2 Tie cotton threads to the leaf stalks, hang them from the arm of the balance and weigh them.
3 Tie the leaves on to a suitable support or on to a line across the room.
4 After a few hours re-weigh.

RESULTS

	Vaseline on upper surface	Vaseline on lower surface
Before	g	g
After	g	g
Gain or loss in weight	g	g

CONCLUSION

The leaves with vaseline on upper surface...............weight while those with vaseline on lower surface...............weight. Therefore, water was.................through the lower surface which contains............, only found on this surface of the leaves.

EXPERIMENT 11

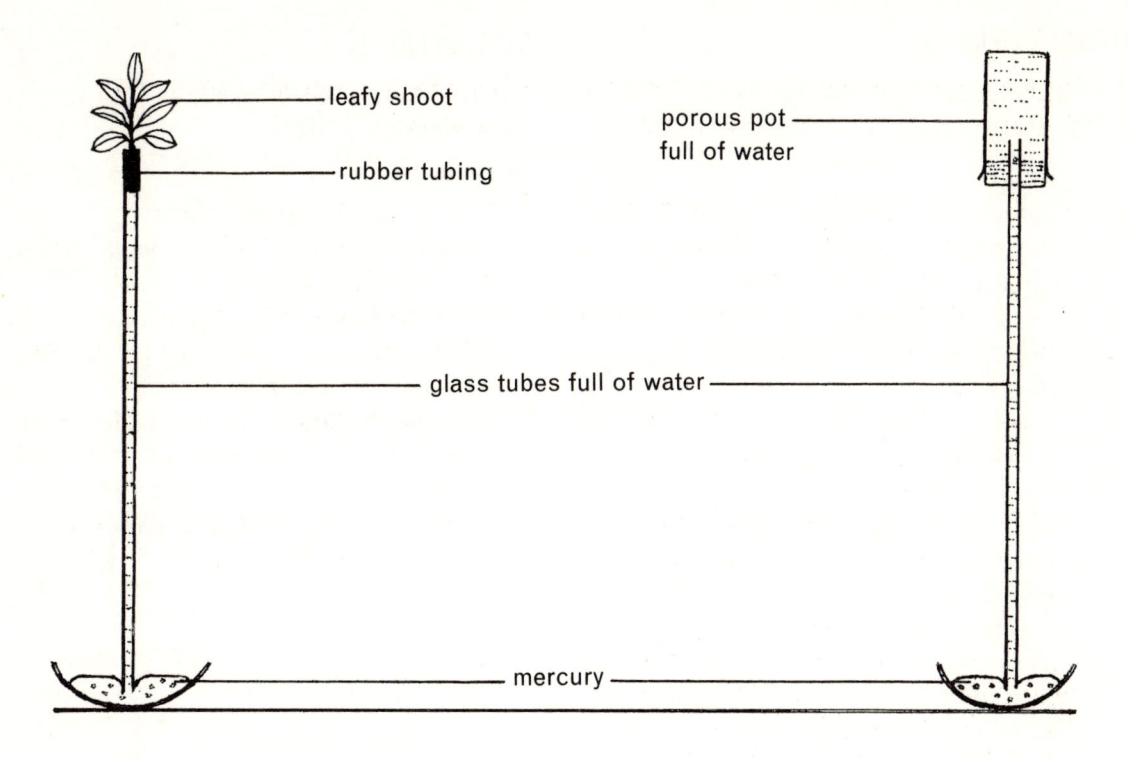

Does the water escaping from the leaves of a plant help to draw water up the stem?

METHOD

1. Cut off a leafy shoot of privet under water.
2. Take a 100 mm length of rubber pressure tubing of the same internal diameter as the total diameter of the twig and push into the tubing a cork borer of slightly larger size than the stem.
3. Under water place the cut end of the twig into the open end of the cork borer and, holding the rubber tubing tightly, carefully withdraw the borer to allow the tubing to hold the stem for about 50 mm.
4. To the open end of the tubing attach a 600 mm length of glass tubing full of water. Make sure no air bubbles are present in the tubing.
5. Place your finger over the open end of the glass tubing and then place this end under some mercury in an evaporating dish.
6. Support the apparatus with a retort stand.
7. Set up an identical apparatus but with a porous pot full of water attached to the glass tubing instead of the leafy shoot.
8. After a few hours do you notice any change in the level of the mercury in the glass tubes?

What conclusions can you draw from your observations?

EXPERIMENT 12

What is the nature of the surface of the skin and what is its reaction to a volatile liquid?

METHOD

1. Examine the skin on the back of your hand under a lens. Is the surface of the skin smooth? If not, how would you describe its appearance? What can you see growing up from the skin?
2. Press your fingers on a cold glass or a polished surface. What can you now see on this surface? What is this due to?
3. Press the fingers of one of your hands on the back of the other. Does the skin change colour? Does the colour return when the pressure is removed? How do you explain these observations?
4. Place one or two drops of ether or chloroform on the back of your hand. Does it feel hot or cold? Wave your hand about in the air, is the effect increased? How do you account for these observations?

EXPERIMENT 13

Are all parts of the hand equally sensitive to heat and cold?

METHOD

1. Dip the blunt end of a metal rod in very hot water. Test your fingers, the palm, and the back of your hand with this rod and put a spot of red ink where you can most detect the heat.
2. Repeat this procedure with a rod dipped in ice water and put spots of blue ink where you can most detect cold.

Are the spots of heat and cold in the same place?

EXPERIMENT 14

Are our hands a reliable guide to the temperature of water?

METHOD

1. Prepare three beakers of water, one containing hot water, one ice-cold water, and one warm water.
2. Place one finger in the hot water and the other in the cold water. Leave them in the water for a minute or so.
3. Then place both fingers in the warm water. Does it feel the same to both fingers?

Would you say that the hands are a reliable guide to the temperature of the water?

EXPERIMENT 15

To find out if all parts of the hand are equally sensitive to touch

METHOD

1. Close your eyes and allow your partner to touch the skin on different parts of the hand with a pair of dividers.
2. Get him to mark on your skin the spots where you can only distinguish one point and not two.

Are these 'touch spots' evenly distributed over the skin? If not, where are they mainly found? How far apart are the dividers when you can only distinguish one point?

CHAPTER 7

The structure, function and circulation of blood in vertebrates

EXPERIMENT 1
What is the blood composed of?

METHOD A
1. Tie a piece of cotton or string tightly round the last joint of the index finger. Sterilise this area by wiping it over with alcohol.
2. Sterilise a needle by passing it two or three times through a flame.
3. Bend the end of the finger and prick the skin just below the nail. What happens? What is the purpose of the tightly tied cotton or string?
4. When you have obtained a drop of blood, quickly smear it across a warm glass slide. Place a cover slip across a thin part of the smear.
5. Look at your blood under the high power of the microscope.

OBSERVATIONS
Can you see many tiny round pale orange coloured cells? What are they? Can you see any other cells which are not round? Are there many of them compared with the round ones? What are these other cells called? Are all these cells moving or are they stationary? If they are moving, how would you describe the movement?

CONCLUSION
Our blood contains .... different types of cell which are .... in a fluid.

METHOD B
1. Obtain a drop of blood as before.
2. Put the drop on the end of a perfectly clean slide and put the flat edge of another clean slide in the drop.
3. Draw the drop out by pulling the second slide along the surface of the first. You have thus made a smear or blood film.
4. Dry this smear quickly by waving it in warm air, e.g. over the radiator.
5. Put a drop of Leishmann's stain on the smear, and, by gently rocking the slide, mix with it a drop of distilled water.
6. After a couple of minutes wash the slide under the tap to remove the stain. Dry the slide by waving it in the air.
7. Look at the slide under the microscope.

OBSERVATIONS
Can you see any cells which are transparent except for a reddish purple centre. Are there many of these? In what ways are they different from most of the cells you can see? What are these stained cells? What is the function of these different cells?

EXPERIMENT 2
To discover under what conditions haemoglobin changes colour

METHOD
1. Obtain a few drops of fresh blood as described in Experiment 1, and transfer them as quickly as possible to a test tube.

2 Add to the test tube for every drop of blood twenty drops of water.
3 Put your finger over the end of the test tube and shake it vigorously. What is the colour of the liquid?
4 Add a few crystals of sodium hyposulphite. Is there any change in the colour of the liquid? If so, what change?
5 Again shake the liquid for a few minutes. Is there any change in the colour of the liquid? How do you account for the observations you have made?

EXPERIMENT 3

To discover what happens to a drop of blood when it is exposed to the air

METHOD

1 Place a drop of blood, obtained as in Experiment 1, on a slide.
2 Look at this drop through the microscope at intervals of five minutes, taking great care to keep the slide perfectly still.

OBSERVATIONS

After twenty minutes or so can you see any difference in the arrangement of the red blood cells? If so, how would you describe this in everyday terms? Can you see any structures appearing which were not in the blood before? If so, what are they and where do you think they come from? What process would you say was taking place in the drop of blood?

EXPERIMENT 4

In which direction is the blood flowing through the veins?

METHOD

1 Roll up your sleeve to just above the elbow.
2 Tie a handkerchief fairly tightly around your arm just above the elbow. Clench the fist.
3 Closely observe the veins. Do you notice any change taking place in them?
4 With your finger, stroke the veins downwards.

OBSERVATIONS

Do any small swellings appear? If so, what could they be due to? What happens to these swellings if you stroke the veins upwards? What does this tell you about the path of blood through the veins?

EXPERIMENT 5

To discover how blood circulates through the tail or external gills of a tadpole

METHOD

1 Place a tadpole which has just hatched out in a shallow watch glass of water.
2 Add to the water one drop of chloroform. What is the effect of this?
3 Look at the external gills or the thin part of the tail under the microscope.

OBSERVATIONS

Can you see any very small tubes? If so, what do you think they are? Can you see any small bodies passing along these tubes? Do they travel in all directions or just one? What do you think they might be? Make drawings of your observations.

EXPERIMENT 6

To find out the internal structure of a sheep's heart

METHOD

1 With a scalpel make a deep cut on either side of the lower pointed end of the heart, and with a pair of scissors carefully enlarge and widen the cuts you have

made. Are the two cavities that you have cut into connected to each other? Is there any difference in the thickness of the walls of the two sides of the heart? If so, how do you account for this? What are these two cavities that you have exposed called? Can you see any thin white cords at the top of each cavity? If so, to what are they attached? What is the function of these cords? Can you find any more cavities in the heart? If so, how many and where are they in relation to the two that you have already found? Are they larger or smaller than these? How do you think the blood passes from one cavity to another?

2 With a probe or your finger can you find any tube leading out from the upper cavity on either side? If so, cut into one of them near its base. Can you see any 'flaps' of tissue attached to these walls? If so, pour water down on to them. What happens? What do you think their function is? Are there any other tubes running into or out of the heart? If so, what are they? With a probe try and trace where they go to or come from Work out how the blood enters, passes through and leaves the heart and draw a diagram to illustrate this.

EXPERIMENT 7

Does the pulse beat vary after exercise?

METHOD

1 Find your pulse beat by placing your right index finger just below the hard bone in the left wrist with the palm upwards.
2 When you are quite sure you can feel it beating, count the number of beats in one minute.
3 Repeat this twice and from your results calculate the average beat.
4 Repeat this when you are standing up or lying down. Record your results.
5 Carry out some exercise for a few minutes, e.g. run twice round the school, or up and down the stairs twice. Measure your pulse rate every two minutes until it returns to normal and record your results in the form of a graph of rate against time.

RESULTS

Pulse rate/min	1	2	Average
Sitting . .			
Standing . .			
Lying . .			
After exercise			

How do you account for your results?

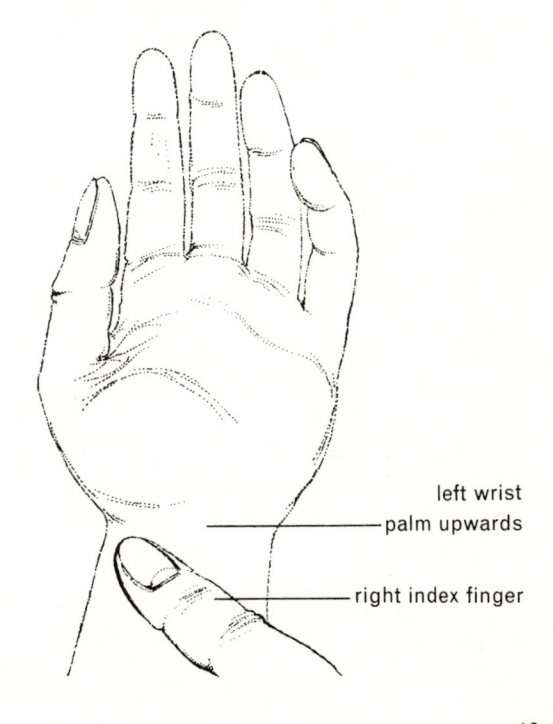

left wrist
palm upwards
right index finger

EXPERIMENT 8

How quickly will snails come out of hibernation?

METHOD
1 Weigh the snail carefully.
2 Place the snail in a beaker of water kept at 15° C, and time how long it takes for the snail to appear (with its head fully out). What happens to the epiphragm? Is there more than one?

RESULTS
Compare your results with that of the rest of the class by constructing a graph of weight against the time taken for the snail to appear.

CONCLUSION
Does the weight of the animal have any effect on the time it takes to appear?

EXPERIMENT 9

To discover what external conditions cause snails to hibernate

METHOD
1 Collect a large number of active snails.
2 Obtain four jam jars with perforated lids.
3 Place five snails in one jar and keep it dark, warm (near a radiator), and dry. Call this jar A.
4 Put another five snails in another jar and place it in a refrigerator, keeping it dry. This is jar B.
5 Repeat with jar C and place this in the dark, keeping it warm and also damp by placing a moistened piece of cotton wool in the jar.
6 Repeat with jar D and place this in the warm. Keep it dry but well lit all the time.
7 Examine the jars each day and note down when the snails have made epiphragms.

RESULTS

Conditions	Time taken to form epiphragms
A. Dark, warm, dry	
B. Dark, cold, dry	
C. Dark, warm, wet	
D. Light, warm, dry	

CONCLUSION
The conditions which lead to the formation of epiphragms are ....
....

CHAPTER 8

Animal movement and its control

EXPERIMENT 1
How does an earthworm move?

METHOD
1 Carefully wash off the soil from the worm.
2 Slip the worm into a 600 mm length of glass tubing which is just slightly larger in diameter than the worm.
3 Observe how it moves. How would you describe its movement? Are all parts of the body of equal diameter as it moves? How long does the worm take to travel along the tube? Can it move along the tube when this is placed vertically?
4 Remove the worm from the tubing and lightly run your finger along the light-coloured side of the body. Can you feel anything? Look closely at this side with a lens. Can you see any structures which might explain what you felt as you ran your finger along it? What would be a suitable name for these? How do you think they help the worm to move?
5 Place the worm on a clean sheet of paper. As it moves, put your ear close to it. Can you hear anything? If so, what is it caused by?

EXPERIMENT 2
To discover how a snail moves

METHOD
1 Place an active snail on a slightly moistened glass plate.
2 Support the glass plate a few inches above a flat mirror.
3 As the animal moves along look closely at the underside of the foot in the mirror.

Can you see any changes taking place in the foot? Does the foot change shape like the earthworm? What is the purpose of the slime produced by the foot? Does the shell rock about during movement? From your observations how do you explain how the snail moves?

EXPERIMENT 3
How is the lower arm moved?

METHOD
1 Roll up your sleeve to expose the upper arm.
2 Place your finger tips on the uppermost muscle (the biceps muscle) of your upper arm and with fist clenched, raise the lower arm to bring the fist up to the shoulder. What changes in shape take place in this muscle as you do this? Is the muscle firm or soft? Place your fingers in the crook of the elbow. What can you feel?
3 Place the finger tips on the underneath muscle (the triceps muscle) of the upper arm. What changes take place in this muscle as you repeat the experiment? Are both muscles contracted at the same time?

How do you account for your observations?

EXPERIMENT 4

To discover what happens to a muscle if it is given a small electric shock

METHOD

1 Obtain a battery or number of batteries in series, that will produce 10 volts.
2 Connect the terminals to separate pieces of wire and place the bared end of one of these on the spinal cord of a recently killed frog.
3 Place the other wire on the skin of the hind leg—what happens?

How do you explain your observations?

EXPERIMENT 5

Testing our reflexes

METHOD

1 Sit down with the right leg crossed over the left allowing the right one to hang freely. (This stretches the 'femoral muscle' which has a tendon running to the knee cap. The knee cap is connected to the top of the tibia by another tendon.)
2 With the edge of your hand give a sharp tap just below the knee cap. Describe what happens. Why is the reaction so quick?

Of what value are reflexes to the body?

CHAPTER 9

The senses

EXPERIMENT 1

To find out the action of the iris under different intensities of light

METHOD

1. First look at your partner's eye. What colour are the eyeball, the iris and the pupil? Is the pupil large or small?
2. Get your partner to cover his eye for a minute or so, and when he removes the cover quickly look at the size of the pupil. Has it changed since your first observation?
3. How do these changes in the pupil take place? What is their value?

EXPERIMENT 2

How far from the eye must an object be held for its image to fall on the blind spot?

METHOD

1. Mark in the middle of a piece of paper a cross and a spot 100 mm apart, with the cross on the left.
2. Close your left eye and holding the paper at arm's length in front of your right eye concentrate on the cross and slowly bring the paper towards you. While you do this, are you aware of the spot all the time? If not, at what distance from the eye does it disappear?
3. At this distance move the paper from left to right. What happens to the spot?
4. Repeat the experiment with a cross and a spot only 50 mm apart. Does the spot still disappear at a certain point? If so, what is this distance?
5. Draw a diagram to illustrate your observations and conclusions.

EXPERIMENT 3

Why do we need two eyes?

METHOD A

1. With your left eye closed, hold a pencil at arm's length in the right hand, and line it up with the wall of a distant building.
2. Open your left eye and at the same time cover the right eye with the left hand. What does the pencil appear to do?
3. Repeat the experiment with the pencil much closer to the eye this time. What do you observe? How do you account for your observations?

METHOD B

1. Hold a pencil, point upwards, in each hand at arm's length. Can you make the points touch?
2. Now close the left eye. Can you now make the points touch?

What conclusions can you draw from this experiment?

EXPERIMENT 4

To find out if the sensitivity to hearing of members of the class varies

METHOD

1. Start out of earshot of a clock and walk slowly towards it until you can just hear it ticking.

2 At this point move slightly backwards or forwards to finalise your position. Measure this distance.
3 Make a table of the class results. Is everyone in the class equally sensitive to the sound of the clock?

EXPERIMENT 5

To find out if sea anemones have a sense of taste

METHOD

1 Place an anemone in a bowl of sea water.
2 Make up 4 compact masses of cotton wool. Tie each mass to a piece of cotton thread and soak them thoroughly in sea water.
3 Mix separately equal quantities of flour, saliva and sugar with sea water, and soak a mass of cotton wool in each.
4 Gently lower the untreated mass of cotton wool on to the anemone tentacles and observe carefully their action. Do all the tentacles respond to this? Is the cotton wool taken into the mouth? If not, what happens to it?
5 Repeat this with the treated masses and each time observe carefully what happens. Squirt fresh sea water on to the anemone after each treatment and allow it to rest for a few minutes. Does the anemone appear to be going to eat any of the treated masses? If so, remove the mass before it is taken into the mouth.
6 Finally, feed the anemone with a small piece of meat, e.g. a worm, mussel, limpet or winkle. Do all the tentacles convey food to the mouth?

Do you think anemones have a sense of taste?

EXPERIMENT 6

To find out if earthworms have a sense of taste

METHOD

1 Obtain a number of flower pots filled with a good loam soil and place earthworms in each.
2 Leave them for a few days while the worms settle in, and then place on the surface a number of small pieces of food, e.g. bread, potato, carrot, chocolate, meat, cheese, and leaves of trees.
3 Place the pots in a dark place and keep them slightly damp.
4 Look at the pots after a few days. Have any of the pieces of food been drawn into the soil? If so, carefully remove these. Have the ends become changed in any way? Have any of the leaves been drawn into the soil? If so, were they all pulled in by the same end?

From your observations would you say that a worm has a sense of taste?

EXPERIMENT 7

Do our tongues have areas which are more sensitive to some tastes than others?

METHOD

1 Take a small piece of blotting paper in a pair of tweezers or forceps, dip it into solution A and touch it on the tip, front, sides, and back of your tongue. Are there any areas of the tongue which are particularly sensitive to the taste of this solution? If so, what are they?
2 Repeat this procedure with a fresh piece of paper for solutions B, C, and D.
3 Draw a diagram of your tongue and write on it the areas which are most sensitive to the different tastes.

CHAPTER 10

Chemical co-ordination in plants and animals

EXPERIMENT 1

To find out the direction of growth of broad bean roots placed vertically and horizontally

METHOD

1 Obtain three germinating broad beans with straight radicles about 25 mm long.
2 Place a piece of cork mat covered with blotting paper in a jam jar containing about 100 mm of water.
3 Through the fleshy part of the seeds, pin one bean to the cork with the radicle vertical, and another with the radicle horizontal.
4 Cut off the radicle tip of the third bean and pin this one to the cork with the radicle horizontal.

OBSERVATION

Has any change in the position of the radicles taken place after a few days? If so, in which ones?

CONCLUSION

How do you account for your observations?

EXPERIMENT 2

In what region of a broad bean radicle does bending occur?

METHOD

1 Obtain a large cork and push into its broad end four pins to form legs.
2 Drape and pin down over the cork a piece of blotting paper.
3 With Indian ink, mark off divisions of 1 mm on the straight radicle of a bean.
4 Pin the bean on to the top of the cork and place it in a jam jar containing 15 mm of water. Put a sheet of glass over the jar and keep it in a dark place.

OBSERVATION

Look at the bean after a day or so. Are the divisions along the radicle still of equal width? If they are not, where have the divisions become wider?

CONCLUSION

What conclusions can you draw from your observations?

EXPERIMENT 3

To find out if oat seedlings respond to light coming from one direction

METHOD

Obtain five sets of oat seedlings which have been grown in the dark and label them A, B, C, D, and E respectively. Treat them as follows:

Set A Keep these in the dark.

Set B Place these in a cardboard box which has a small window cut in it on one side. Expose this window to light coming from one direction by placing the box on a window sill for an hour or by shining a bench lamp on it.

Set C Cover the tips of the seedlings with very small aluminium foil caps and then expose to light coming from one direction as with Set B.

Set D Expose the seedlings to light coming from one direction as with Set B, and then cover tips with foil as before.

Set E Cut off the first two mm of the tips of the seedlings and then expose to light from one direction as before.

Draw a diagram of each set and then return them to the dark. After a day or so what differences, if any, can be seen in the sets? Draw diagrams to show this. Has the exposure to light coming from one direction had any effect on the direction of growth of the seedlings? What is the effect of cutting off the tips? What conclusions can you draw from your observations?

EXPERIMENT 4

To find out how roots respond to the presence of water

METHOD

1 Take a small empty flower pot and stop up the drainage hole with a rubber stopper.
2 Place it in the centre of a deep pie dish.
3 Half fill the dish round the pot with damp sand.
4 Place some soaked peas on the sand at varying distances from the flower pot and cover with damp sand to the edge of the dish.
5 Fill the flower pot with water and for the next two weeks keep this filled but do not water the sand.
6 Set up an identical apparatus but keep the flower pot dry and water the sand regularly and evenly.
7 After two weeks carefully dig out the seedlings from the two sets of apparatus.

Can you see any difference in the direction of growth of the roots of the two sets of seedlings?

Make drawings to show any differences.

What conclusions can you draw from your observations?

CHAPTER 11

Bacteria and moulds — agents of decay and disease

EXPERIMENT 1

To discover what happens when food is left uncovered in the air for a short length of time

METHOD

1 Obtain a piece of moist bread, a piece of half-cooked potato, a slice of orange (or pear), a piece of cheese, a portion of jam and a piece of banana skin.

2 Expose these to the air for half an hour.

3 Place them separately on the top of paste jars which are standing in saucers of water. Cover with sterilised jam jars.

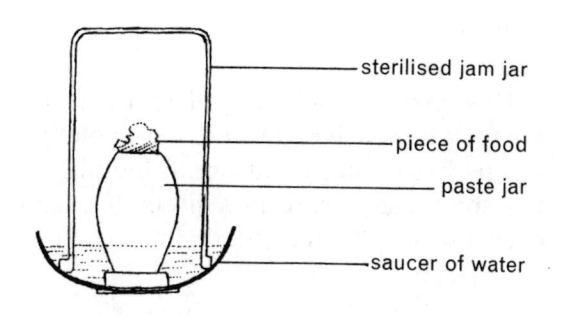

- sterilised jam jar
- piece of food
- paste jar
- saucer of water

4 Keep the jars in a warm place, e.g. near a radiator

5 Make drawings of your experiments as you set them up.

6 Look at the foods after a week or so.

OBSERVATION

Has any change in the foods taken place? If so, what is this change due to?

Make drawings of the pieces of food if a change has taken place.

EXPERIMENT 2

To obtain some spores and observe their growth under laboratory conditions

METHOD

1 Sterilise a paint brush by dipping it in boiling water for a few seconds.
2 Draw this brush across the surface of a growing mould.
3 Quickly lift up the lid of a dish of potato jelly, draw the brush across the surface of the jelly and quickly replace the lid. Keep the dish in a warm place.
4 Examine the dish at intervals with a lens.

OBSERVATIONS

Make a series of drawings to show any changes taking place.

CONCLUSION

What conclusion can you draw from your observations?

EXPERIMENT 3

What is the effect of adding yeast to a sugar solution?

METHOD

1. Add a pinch of yeast (or a few ml of yeast suspension) to 100 ml of 10% sugar solution in a conical flask.
2. Connect a delivery tube to the flask and connect the other end of the tube by a small piece of rubber tubing to a corked test tube of lime water.
3. Place the apparatus in a warm place for an hour or so.

OBSERVATION

What happens to the liquid in the flask?
What happens to the lime water in the test tube?
How do you account for these changes?

EXPERIMENT 4

To discover where bacteria are to be found

METHOD

1. Obtain dishes of sterile agar jelly and treat as follows. (In each case raise the lid from one edge, and replace it as soon as possible.)
   - (a) Sprinkle fresh soil over the jelly in the first dish. Allow some to stick to the jelly and gently shake off the remainder. Replace the lid.
   - (b) Open another dish to the air in the Science Room for five minutes, and then replace the lid.
   - (c) Do the same in the air of the playground. Replace the lid.
   - (d) Do the same in the air of the lavatory. Replace the lid.
   - (e) Using a sterile needle remove some dirt from underneath a fingernail and then draw the needle carefully across the surface of the jelly. Replace the lid.
   - (f) Using a sterile needle carefully remove food from between the teeth and draw the needle carefully across the surface of the jelly. Replace the lid.
   - (g) Slowly draw the unsharpened end of your pencil across the surface of the jelly.
   - (h) Open the dish and with a pair of tweezers gently place a coin or some paper money on the jelly. Leave for five minutes with the lid on, remove it and replace the lid.
2. Place all the treated dishes in a warm place and after a day or so look at them again.

OBSERVATIONS

How many colonies of different bacteria are found on each plate? Are some colonies of one bacterium found on all the plates? Do some colonies form a liquid-like film over the jelly? Draw diagrams to show your results.

EXPERIMENT 5

To discover what happens to bacteria in milk when it is heated at different temperatures

METHOD
1. Take four test tubes and into each place 10 ml of milk and a pinch of fresh soil. Close the end of each test tube with a plug of singed cotton wool. Label the tubes A, B, C, and D.
2. Heat test tube A for 4 minutes at 100° C, test tube B for 30 minutes at 40° C, and test tube C for 60 minutes at 25° C. Leave test tube D untouched.
3. Allow the tubes to cool and examine them at intervals.
   Why is test tube D left untouched?

OBSERVATIONS

Is there any change in the condition of the milk in the four test tubes? If there is change, in which test tube does it take place first?

CONCLUSION

What conclusions can you draw from your results?

EXPERIMENT 6

To repeat Pasteur's experiment which proved that bacteria in the air cause decay

METHOD
1. Boil some chopped beetroot in water. After boiling strain off the liquid.
2. Place some of this liquid in flasks A and B, each fitted with stopper and tube (see diagram below).
3. Sterilise these by either steaming in a saucepan for at least 30 minutes or in a pressure cooker at 10 lb. ($= 68 \cdot 94$ kN/m²) pressure for 15 minutes.
4. Leave the flasks in a warm place.
5. Examine daily.

OBSERVATIONS

In which flask does the liquid go cloudy? What is this cloudiness due to?

CONCLUSION

Why does the liquid go cloudy in one flask and not the other?

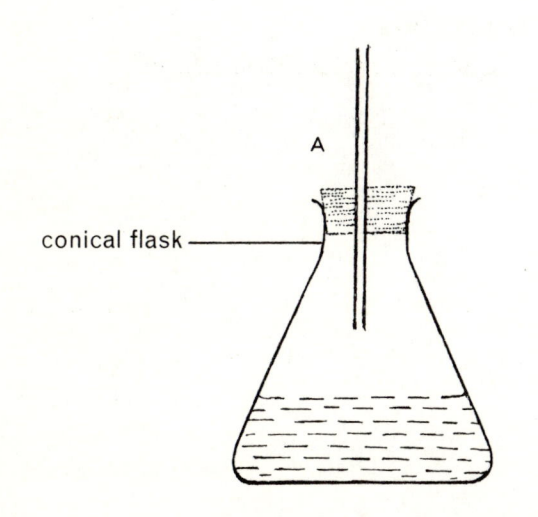

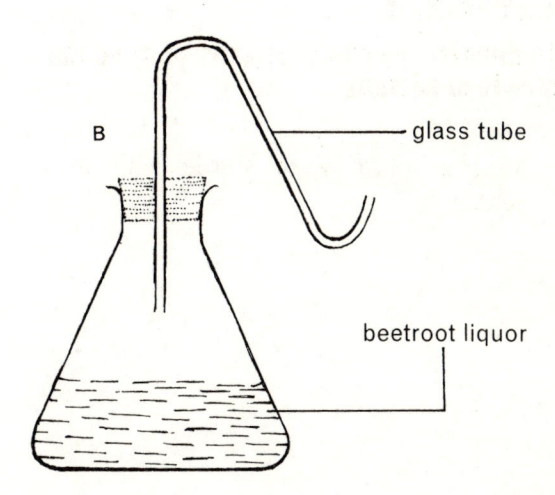

EXPERIMENT 7

To demonstrate how bacteria can be transferred from place to place

METHOD

1. Take five dishes of sterile jelly, label (*a*) to (*e*), and test as follows:
   - (*a*) Leave untouched.
   - (*b*) Cough into the jelly and replace the lid.
   - (*c*) Allow a fly to walk over the jelly, for a minute or so, with the lid on. Release the fly and replace the lid.
   - (*d*) Gently press your fingers on the surface of the jelly. Replace the lid.
   - (*e*) Wet your hands thoroughly and again gently press your fingers on the surface of the jelly. Replace the lid.
2. Place the dishes in a warm place.

OBSERVATION

After a day or so has anything appeared in the dishes? What is the purpose of dish (*a*).

Draw diagrams to show your results.

CONCLUSION

From your observations say in what way bacteria can be transmitted.

EXPERIMENT 8

To discover the effects of antiseptics on the growth of bacteria

METHOD

1. Melt a quantity of sterile jelly in a beaker.
2. To 6 separate 10 ml quantities of melted jelly add:
   - (*a*) 1 drop of Dettol.
   - (*b*) 1 drop of 50% Dettol.
   - (*c*) 1 drop of very weak Dettol.
   - (*d*) 1 drop of Milton.
   - (*e*) 1 drop of Savlon.
   - (*f*) 1 drop of iodine.

   In each case pour the treated jelly into sterile petri dishes or sterile test tubes.
3. Replace lid or cotton wool plug.
4. Allow the jelly to set. If a tube is used, allow the jelly to set sloping.
5. With a sterile needle draw a suspension of bacteria across the jelly in the tubes (the suspension is prepared by mixing bacteria from previous plates in a little distilled water on a glass slide). Replace lids or plugs.
6. Set up one dish or tube without treating the jelly.
7. Place all the dishes or tubes in a warm place. Observe at intervals.

OBSERVATION

Is any growth of bacteria noticed? If so, how long does it take to appear, and in which dish or tube does it happen?

What effect do the liquids appear to have? Are they all equally effective?

CONCLUSION

Can you explain their action?

EXPERIMENT 9

To discover various ways in which food can be preserved

METHOD

1. Melt a quantity of sterile jelly in a beaker.
2. To three separate 10 ml portions of melted jelly add:
   (*a*) A few ml of strong salt solution.
   (*b*) A few ml of vinegar.
   (*c*) A few ml of strong sugar solution.
3. Pour the treated portions into three separate petri dishes, replace lids and allow jelly to set.
4. Expose the prepared dishes, and one other untreated dish, to the air in the room for one hour. Replace the lids.
5. Place the untreated dish in the refrigerator, and the others (including another unexposed and untreated dish) in a warm place. Observe at intervals.

OBSERVATIONS

Do any of the above solutions slow down or prevent the growth of bacteria or moulds? What effect has a low temperature on this growth? What is the purpose of the untreated and unexposed dish?

FURTHER CONSIDERATION

In what ways can we use any of the above solutions for preserving our food?

EXPERIMENT 10

Do root nodules contain bacteria?

METHOD

1. Obtain a plant bearing root nodules, e.g. clover, lupin, pea or bean.
2. Wash the roots thoroughly.
3. Cut off a nodule and crush it in a drop of water on a clean glass slide with the sterilised end of a metal scalpel.
4. Spread the crushed material along the slide by drawing the edge of another slide over it.
5. Pass the slide two or three times quickly through a flame.
6. Place on the slide a drop of Indian ink diluted with a drop of water and allow the diluted ink to cover all the material.
7. Dry the slide by waving it in the air and look at it under the microscope.

OBSERVATION

Can you see any light objects against a dark background? If so, what shape are they?

FURTHER CONSIDERATION

Of what value to the plant are these bacteria in the root nodules? Can you suggest ways in which these plants might be useful in agriculture?

CHAPTER 12

Field studies

EXPERIMENT 1

To compare the rate of fall of sycamore fruits with those of ash

METHOD

1. Obtain a large number of sycamore and ash fruits.
2. Weigh a number of each of these and time how long each takes to fall a known distance. (This might be from one's hand at the top of the school stairs.)

OBSERVATION

How do they fall? Do they fall straight down? Is there any connection between their weight and the time taken to reach the floor? Find the average rate of fall for all the sycamore fruits, and then the ash fruits. Is there any difference between them? How does their shape aid their dispersal from the parent tree? Which of them would carry the further in a strong wind?

EXPERIMENT 2

To discover how far from the parent plant poppy and delphinium seeds are scattered

METHOD

1. Obtain a number of dried stems of poppy and delphinium still bearing the intact fruits.
2. Fix these firmly in a clamp on a retort stand placed on the bench.
3. Place a large number of sheets of newspaper on the bench and, holding the base of the stand firmly, bend the dry stems back a little and then release them. How far back must the stem be bent for any seeds to be thrown out?
4. When this distance has been discovered, repeat the setting up and on releasing the stalk make a count of seeds found at intervals of 150 mm away from the stand. How many are found in each zone? What is the furthest distance that the seeds are thrown? Are poppy seeds thrown further than those of delphinium,

or vice-versa? Find the average weight of a number of seeds of each plant. Is there any difference in this weight and, if so, has it any bearing on the results of your experiment?

CONCLUSION

Which of the two types of capsule do you consider most efficient in the dispersal of seeds?

EXPERIMENT 3

To discover how fast fluffy fruits and seeds fall

METHOD

1. Obtain a large number of fluffy fruits and seeds, e.g. dandelion, thistle fruits, rose-bay willow herb seeds, and a long 50 mm wide glass tube.
2. Measure the length of the tube and time how long the fruits and seeds take to reach the bottom.
3. Find the average weight of a large number of these fruits and seeds. Has this weight any connection with the results obtained?

CONCLUSION

How does the shape of these fruits and seeds aid their dispersal?

EXPERIMENT 4

To discover how seeds are dispersed from pods

METHOD

1. Obtain fruits of everlasting pea, geranium or rose campion.
2. Place these alternatively in steam from a kettle or similar container, and close to a bench lamp. Allow 10 minutes in each situation. What happens on doing this? From this, can you suggest how the seeds are dispersed?
3. Place some pods of everlasting pea near a source of heat, e.g. a bench lamp. If possible, record how far from the pod the seeds are thrown.

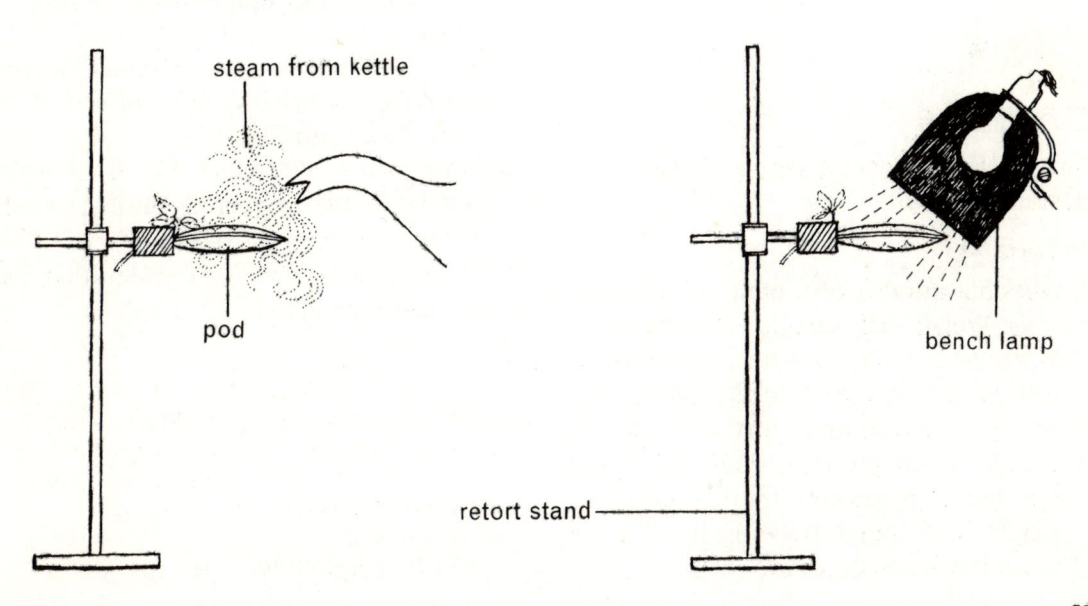

EXPERIMENT 5
To find out what soil is composed of

METHOD

1 Obtain a number of samples of soil from a variety of sites.
2 Place each sample in a tall beaker or gas jar, add distilled water to within 25 mm of the top, and placing one hand firmly over the open end shake up the sample thoroughly. When the water was first added, what did you observe?
3 Allow the soil to settle. What particles settle first and what particles last? What do you find floating on the surface? Are the same quantities of particles of different size found in each sample? If not, measure the respective amounts and make diagrams of each sample to show their distribution.
4 Evaporate some of the water to dryness. What is left? Where has this come from?

EXPERIMENT 6
How much water does a sample of soil contain?

METHOD

1 Obtain a number of samples of different soils. Weigh each sample carefully.
2 Place the sample on a tin lid and allow it to dry out in the air of the laboratory, or dry it by heating at just over 100° C.
3 Weigh the sample when cool and repeat the above procedure until a constant weight is obtained; showing that all the water has been removed.

RESULTS

Wt. of soil before drying = ... g
 ,, ,, ,, after ,, = ... g
∴ Sample contained ...... g of water.
∴ Sample contained
$\left(\dfrac{\text{wt. of water}}{\text{wt. of soil before drying}} \times 100 \right)$ % of water.

How does this compare with other samples?

EXPERIMENT 7
Which type of soil holds most water?

METHOD

(*Try to obtain samples of sandy and clay soils.*)

1 Stand a number of filter funnels in 100 ml measuring cylinders.
2 In each of the funnels place a plug of cotton wool.
3 Half fill the funnels with equal weights of previously dried samples of soil. Press these down firmly.
4 Pour into each funnel 50 ml of water and leave the apparatus until no more water is draining through.
5 Measure the volume of water that has drained through.

RESULTS

	Sample A	B	C	(etc.)
Amt. of water poured on	50 ml	50 ml	50 ml	
,, ,, ,, drained through	... ml	... ml	... ml	
∴ amt. of water retained by sample	... ml	... ml	... ml	

Which sample holds the most water?

EXPERIMENT 8
How much air does a sample of soil contain?

METHOD

1. Obtain a medium sized baked bean tin or similar tin which has a sharp rim.
2. Calculate the volume of the tin with water and a measuring cylinder.
3. Make a few holes in the bottom of the tin with a nail.
4. Push the tin down into the soil until the bottom is level with the surface of soil.
5. Scrape away the soil from around the tin and dig the tin out without disturbing the soil inside. The soil in the tin is then made level with the rim of the tin.
6. Obtain a large measuring cylinder containing 200 ml of water.
7. Tip the soil from the tin into the measuring cylinder.
8. Stir the soil and water.
9. Read off the final level of water in the cylinder.

RESULTS

2nd reading of measuring cylinder (water and solid part of soil)	=	ml
1st reading of measuring cylinder (water without solid part of soil)	=	ml
Difference = volume of solid part of soil	=	ml
Volume of soil taken in tin	=	ml
Volume of solid part of soil	=	ml
∴ Difference = volume of air in soil	=	ml

CONCLUSION

% volume of air in soil =
$$\frac{\text{vol. of air in soil}}{\text{vol. of air in tin}} \times 100.$$

EXPERIMENT 9
To discover some of the physical properties of soils

METHOD

(*a*) Obtain a number of samples of a variety of soils including sand and clay.
Make a 'mud pie' of each sample. Place these on a board and leave to dry. Is the 'pie' crumbly or hard? Has it shrunk or cracked? Is it smooth or hard, rough or dry?

(*b*) Powder some dry clay, put it in a crucible and carefully stand the crucible in a beaker of water. What happens and how do you account for it?

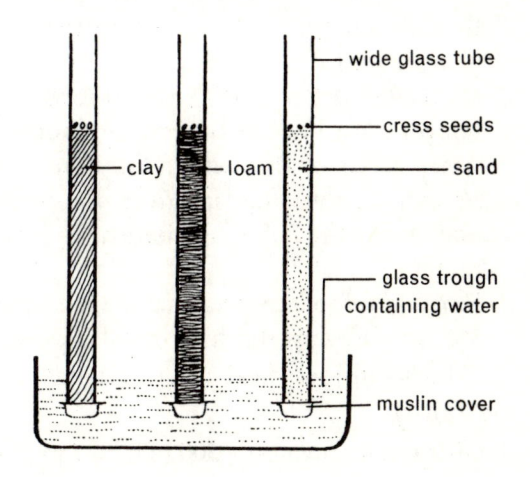

(*c*) What does the following experiment show you? Fill and pack well three glass tubes, each about 600 mm long and 25 mm in diameter, with equal volumes of dry sand, loam and clay. Tap on the glass to see that the soils are evenly packed. Cover the lower ends of the tubes with a piece of muslin firmly tied on. Stand the tubes in about 50 mm of water in a glass trough. Place cress seeds on the top of the soils.

Record the height of the water level in the tubes every minute for the first ten minutes, then every five minutes, and then at longer intervals leaving the apparatus set up for a few days. From your results plot a graph of height of water level against time.

In which tube does the water rise fastest? In which tube does it rise the highest? In which tube do the seeds germinate first?

FURTHER CONSIDERATION

How do you explain your results and what application has this experiment in agriculture?

EXPERIMENT 10

To find out how different coloured soils are affected by heat

METHOD A

1 Fill three metal tins with equal quantities of sand, clay and loam soil. Insert a thermometer in each and place them in a water bath.
2 Heat the water bath and record the temperature in each tin every minute.
3 After a reasonable length of time, say 30 min, remove the tins from the water and record the falling temperatures every minute.
4 Plot a graph of temperature against time. Which soil absorbs heat most readily? Which soil loses heat most readily?

METHOD B

1 Obtain a wooden box, 500 mm × 250 mm × 200 mm deep, and divide it into two compartments. Into these compartments place either a dark soil and a light soil or the same soil with a layer of lime over it in one compartment and a layer of soot over it in the other.
2 In each compartment make holes to take thermometers at 25 mm, 75 mm and 125 mm below the level of the soil.
3 About 50 mm above the surface of the soil in each compartment suspend a 60 W electric lamp.
4 Record temperatures every half hour for a convenient length of time and construct a graph of temperature against time. Which soil absorbs heat most readily?
5 On switching the lamps off, suspend thermometers 25 mm above the surface of the soils. Record temperatures every five minutes and construct a graph of temperature against time. Which soil is 'giving off' (or radiating) most heat?

EXPERIMENT 11

What is the effect of the addition of lime to clay?

METHOD A

1 Make one cup of damp clay and another of clay mixed with lime. Pour a small quantity of water into each cup.

OBSERVATIONS

What happens? What is the effect of adding lime to the clay?

METHOD B

1 Half fill two jam jars with clayey water. Fill one up with distilled water and the other with lime water containing a little slaked lime. Shake both jars thoroughly.

OBSERVATIONS

After 30 min look at both jars. In which jar has the clay settled leaving the water fairly clear? What is the effect of adding the lime to the clay? What application has this in agriculture?

EXPERIMENT 12
How much humus does a soil contain?

METHOD

1. Obtain a sample of dried soil, weigh it, and heat it strongly in a crucible or on a tin lid.

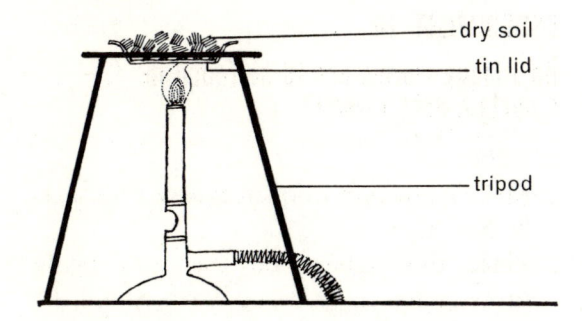

dry soil
tin lid
tripod

What effect does the heating have on the soil? Does it change colour? Can you see any bits of material burning? If so, what do you think they are?

2. Continue heating until the sample falls to a brick-coloured dust. Cool and weigh.
3. Repeat the above procedure until a constant weight is obtained.

RESULTS

Wt. of sample before heating = ___ g
„ „ „ after „ = ___ g
∴ wt. of humus in sample = ___ g

CONCLUSION

% of humus in the sample =

$$\frac{\text{wt. of humus in sample}}{\text{original wt. of sample}} \times 100.$$

EXPERIMENT 13
To find out if a sample of soil contains chalk

METHOD

1. Place a small portion of the sample in a test tube and add a few drops of hydrochloric acid.
2. Do you notice any effervescence or fizzing on adding the acid? What might cause this? Place a drop of lime water held on a glass rod above the fizzing. Does the lime water turn milky? If so, what does this indicate?

EXPERIMENT 14
To determine the degree of acidity of a soil

METHOD

1. Place a small sample of the soil to be tested in a test tube and shake it up with 5 ml of distilled water.
2. Allow the soil to settle and filter the clear liquid off.
3. To this filtrate add two drops of B.D.H. Universal Indicator. What colour is seen?

How acid or alkaline is your sample of soil?

EXPERIMENT 15

To investigate the growth of plants in different solutions

METHOD

1. Sterilize 9 gas jars of 250 ml capacity if you are using cuttings, or 9 crystallizing dishes if you are using duck weed.
2. Cover the sides of these containers with black paper held in position by elastic bands.
3. If you are using gas jars, fit each of them with a cork which has a central hole and a slit through to the hole for inserting the cuttings. These are held in position with cotton wool.
4. Place suitable volumes of the solutions being used in the containers.
5. Place into each gas jar cuttings having the same number of leaves, or into the dishes a definite number of duck weed 'circles', say 20.
6. Place the containers on the laboratory window ledge where there is most light.
7. Record the appearance of the containers as they are set up either by drawings or by means of coloured photographs.
8. Keep the solution in the gas jars at a constant level by adding distilled water. Change the solution in the crystallizing dishes every two days, transferring the duckweed by means of a soft brush.
9. At suitable intervals (for duckweed every day, for cuttings once a week) make the following observations. In the case of the duckweed how many 'circles' are present in each dish? Is there any change in their size or colour? In the case of the cuttings, record their height, number of leaves and branches, length of root system and general appearance. Do they still look healthy?

At the end of the experiment what conclusions can you draw from your observations? What is the effect of a lack of certain mineral salts on plant growth?

EXPERIMENT 16

How many worms are to be found in a certain area of soil?

METHOD

1. Make a 600 mm wooden square with sides 75 mm high.
2. Place this square on the area being investigated and pile up earth round the four sides to seal it up.
3. Slowly pour into the enclosed area 10 to 15 litres of 0·4% formalin.

OBSERVATION

How many earthworms appear? Are they all the same size? Are they all the same colour?

EXPERIMENT 17

What is the effect on the soil of harrowing or hoeing?

METHOD A

1. Half fill two conical flasks with water and place on them filter funnels containing equal volumes of the same soil. The funnels should hold cotton wicks which can pass down into the water. Press the soil down firmly. What is the purpose of the wicks?
2. Leave the flasks for a week, but 'hoe' the surface soil of one flask every day.

OBSERVATION

In which flask has the loss of water been greatest? How do you account for this?

METHOD B

1 Repeat the above experiments but this time instead of 'hoeing' mulch the soil in one flask by placing on top a layer of leaf mould.

OBSERVATION

Again, in which flask has the water loss been greatest?

What is the value of mulching?

EXPERIMENT 18

How far upwards do the spore cases of Pilobolus travel?

METHOD

1 Obtain some Pilobolus fungus which has black spore cases present.
2 Carefully place small amounts in jam jars on moist pieces of cotton wool.
3 Arrange in the jam jars cardboard tubes of varying lengths from 150 mm to 600 mm. Place on the open top of these tubes a piece of glass.
4 Place the sets of apparatus on a well-lit window-sill and observe the glass plates from day to day. On which of the glass plates are black spots visible? What are these black spots?
5 Arrange a further jam jar and tube with all the glass plate covered with black paper except for a 10 mm diameter hole in the centre. Observe this area from day to day. When black spots appear on it, remove the glass plate and look for other spots. Are they to be found all over the plate or just in the unmasked area? How do you account for your observations?

EXPERIMENT 19

How do woodlice behave when given the choice of humid or dry conditions?

1 Obtain a glass trough or round biscuit tin 125 mm deep, and cut out a circular piece of perforated zinc or wire gauze to fit inside the trough.
2 Support this gauze well above two evaporating basins by means of cork or wood supports. Place water in one basin and silica gel in the other.
3 By means of a cardboard fixed to the sides of the trough by sellotape, partition off the two dishes. Do the same thing above the gauze but leave a 10 mm gap at the bottom. Leave for 30 min.
4 Distribute a large number of woodlice evenly over the gauze, cover the whole apparatus with a black cloth or paper, and leave it for 5 min.
5 After 5 min. note the number of animals in each compartment. Then gently redistribute them with a soft paint-brush, leave for another 5 min and count the number again. Record the temperature at each count.

RESULTS

Temp.	Number of woodlice	Time Interval	Number in dry side	Number in humid side	Behaviour of woodlice

How many woodlice should be used? How many results should be taken before making any calculations?

CONCLUSION

How many woodlice prefer the humid side to the dry side?

EXPERIMENT 20

Observation of woodlice in a dry and a damp atmosphere

METHOD

Set up the same apparatus as in Experiment 19, and observe the behaviour of the woodlice in each compartment.

OBSERVATION

How do the woodlice collect in the humid compartment? Do they perform an avoiding reaction at the 'edge' of the dry compartment? In which side are they more active? Is there any tendency for them to huddle together?

EXPERIMENT 21

How do woodlice behave when given the choice of humid or dry conditions at different temperatures?

METHOD

1 Set up two sets of apparatus as in Experiment 19.
2 Place one set of apparatus in a large vessel packed round with ice and salt. Cover the apparatus and make observations as in Experiment 19, taking the temperature at each count.
3 Leave the other set of apparatus at room temperature.

RESULTS

Number of woodlice used	Time interval	Temperature	Humid	Dry

CONCLUSION

Is the behaviour of the woodlice affected by temperature? If so, in what way?

EXPERIMENT 22

How do woodlice behave in relation to light?

METHOD

1 Take an enamel dish and line it with damp blotting paper.
2 Place the woodlice in the dish, put a glass plate over the top, and completely cover one half of the dish with black paper.
3 Illuminate the top of the dish by a bench lamp or place the dish near a well-lit window.
4 Count the woodlice to be found in the light and in the dark at five minute intervals, and redistribute them after each count.

RESULTS

Number of woodlice	Temp.	Time intervals	Number in light	Number in dark	Behaviour of woodlice

CONCLUSION

Are more woodlice to be found in the dark or in the light? How does this relate to the natural habitat of these animals?